THE DEEP STRUCTURE OF BIOLOGY

THE **DEEP STRUCTURE OF BIOLOGY**

Is Convergence Sufficiently
Ubiquitous to Give a
Directional Signal?

Edited by Simon Conway Morris

TEMPLETON PRESS

Templeton Press
300 Conshohocken State Road, Suite 550
West Conshohocken, PA 19428
www.templetonpress.org

The chapters in this collection were originally delivered at the
symposium "Purpose in Evolution," held June 24–26, 2004, at the
Vatican Observatory in Castel Gandolfo, Italy, which was spon-
sored by the John Templeton Foundation as part of the Humble
Approach Initiative.

Designed and typeset by Kachergis Book Design

Library of Congress Cataloging-in-Publication Data
The deep structure of biology : is convergence sufficiently
ubiquitous to give a directional signal? / edited by Simon Conway
Morris.
    p. cm.
  Includes bibliographical references and index.
  ISBN-13: 978-1-59947-138-9 (pbk. : alk. paper)
  ISBN-10: 1-59947-138-8 (pbk. : alk. paper)   1. Convergence
(Biology)   I. Conway Morris, S. (Simon)
  QH373.D33 2008
  576.8—dc22

                        2007045198

Printed in the United States of America

09 10 11 12 13   10 9 8 7 6 5 4 3 2

CONTENTS

INTRODUCTION

To the first approximation, the world is a predictable place; if it weren't, then our space craft would not be able to use the mass of planets as gravitational slingshots to propel them to precise points even in the outer reaches of the solar system, if not beyond. Nor would the lethal gas chlorine and the explosive alkali sodium combine to be sprinkled safely as salt over a lamb chop. Even what are chaotic manifestations—such as the metaphorical flapping of a butterfly's wing in China generating a hurricane that tears across the Caribbean—may overlook the facts that, while the precise reason that a tropical storm first arises will never be known, the turbulent hurricane is a predictable structure and the decadal history of these events also has a given probability.

Science is, therefore, adept at describing and predicting the world around us, but, oddly, this power seems to evaporate when we come to biology. To be sure, the overarching truth of evolution by descent and modification is not in dispute, but, to the first approximation, the processes are regarded as random—think of mutations, or consider the standard view of the historical path as a minefield of the unexpected, as in mass extinctions. Evolution, so the credo runs, is without path or purpose: the end points are indeterminate—think of that most curious of evolutionary flukes, humans.

Or so it would appear. But as the contributors to this volume argue from many different angles, there are certainly aspects of evolution that appear to be constrained, if not predictable. This view largely revolves around the well-known phenomenon of convergence, exemplified by the very similar construction of the camera-eye of the cephalopods (e.g., an octopus) and vertebrates (e.g., a blue whale). Beyond all reasonable doubt—and here we can draw on embryology, comparative anatomy, histology, molecular biology, phylogeny, and the fossil record—the common ancestor of the octopus and blue whale could not possibly have possessed a camera-eye. Each group has independently navigated to the same evolutionary solution, and it is one that not only works very well but has arisen at least five more times, in animals as diverse as snails and, more extraordinarily, jellyfish.

In this volume, many specific examples of convergence are given, in plants and animals, as well as microbes; these are augmented by wider discussions that range from more formal descriptions of convergence to some of the metaphysical implications that necessarily emerge if indeed it transpires that the processes of evolution are far more ordered than is customarily thought. It would certainly be premature to invoke anthropic principles in evolutionary biology, let alone to argue that we can identify general laws and principles such as those that are familiar to physicists and chemists. Yet, at the very least, convergence is a fingerpost in that direction, and this is perhaps most forcibly brought home in terms of the evidence for the independent emergence of intelligence: cognitive maps, mental substrates, and, evidently, mind have all evolved independently from different starting points to strikingly similar destinations. In this context, the comparison between the intelligence of cetaceans and apes is well known, but recent years have also seen startling advances in our understanding of avian intelligence, notably in the corvids, that is, again, intriguingly convergent to the cognitive landscape of the apes. And it is possible that even now our appreciation of intelligence is too zoocentric, as one contributor puts forward an intriguing case for plants as possessing an intelligence.

To use the words *intelligence* and *plant* in the same sentence may well raise eyebrows, and, as all scientists know, the path between inspiration and self-delusion can be painfully narrow. Yet I must emphasize that biology like any science can only progress if the ideas are adventurous, and nobody can complain if a hypothesis fails to survive the rigors of peer review—or alternatively leads to a Nobel. The study of evolutionary convergence seems to mark a potential way forward, but it is certainly not the case that all the chapters here are singing from the same hymn sheet. In fact, very much the reverse is the case: not only is there a proper range of opinion but the very utility of convergence receives hard scrutiny. As an unabashed supporter of convergence—and notwithstanding the fact that, while nobody denies its existence, by no means is everybody persuaded of its importance—I would argue that the biology of the future, one that looks to some general theories of organization, will make today's enthusiasms seem stale and narrow. The selfish gene? An exploded concept that was almost past its sell-by date as soon as it was popularized. Kin selection? Undoubtedly true, but of what wider relevance? Game theory? Ditto! Mass extinctions? Yes, but of what long-term effect?

Irrespective of one's enthusiasm for any of the above or many other fashionable areas of discourse, in no case do I detect any sense that a wider paradigm is being addressed, apart from the uncontroversial and given generalities of neo-Darwinism. And is this not part of our scientific zeitgeist? Is it not popularly supposed that science may be running out of things to do or, more significantly, to say? Now, it may just be true that in cosmology, let alone chemistry or physics, there really is little else to say. Not because we have reached the end of inquiry, far from it, but simply because what appeared to be an almost infinite room of discovery is actually a broom cupboard. Yes, we hear sounds of laughter from adjacent rooms, not to mention the distant murmur from entirely unexplored streets, but the crack of light beneath the door is impossibly narrow and does not even reveal the blocked-up keyhole. Is science apparently running out of things to do simply because scientists hardly know how to tackle the yet-deeper problems that have been revealed? Actually, I doubt this very much, even if areas such as high-energy physics and cosmology have reached an apparent impasse, I am as sure as I can be that this is not true of living systems. At least in the area of biology, one senses that not only have we hardly started, but the simple fact of evolution does little to explain the sheer complexity, fine balance, and potential of living systems. My sense, therefore, is that evolutionary convergence is, at the least, a straw in the wind, pointing to a deeper pattern of biological organization. Metaphorically, does not the ubiquity of convergence point to a map of life, a rugged landscape of almost entirely inaccessible regions that are threaded through by silver roads of vitality? Whether this, in turn, points to a "theory of everything" for biology I rather doubt, but so, too, I am comfortable at the thought of our successors, still many centuries in the future, who will shake their heads at our simplistic thinking as they face new scientific mysteries.

It only remains for me to record first my thanks to the contributors, not only for their uniformly inspiring lectures in the Vatican Observatory, immediately adjacent to the papal palace in Castel Gandolfo, but also their considerable patience in the production process. In this context, I particularly wish to thank the enthusiasm and dedication of Mary Ann Meyers and other members of her team that include Laura Barrett and Natalie Lyons Silver. In Cambridge, I acknowledge with much gratitude the help and organizational skill of Sandra Last and Vivien Brown. So, too, I wish to thank the generosity of both the John Templeton Foundation, for funding both the meeting in Italy and this book, and Father George Coyne, the director of the Vatican Observatory, and other members of his staff. They showed us every kindness, not least in showing us a few of the treasures from their library and recalling that, just as we had a stunning view of the surrounding Campagna, so, too, as we perused the ancient volumes in our hands, we were indeed standing on the shoulders of giants.

Simon Conway Morris
CAMBRIDGE

THE DEEP STRUCTURE OF BIOLOGY

1 CHANCE AND NECESSITY IN EVOLUTION

Richard E. Lenski

Introduction

We humans have long recognized the profound tension that exists in our world between chance and necessity, between things that seem to happen by accident and those that seem inevitable or even purposeful. Democritus said that "everything existing in the universe is the fruit of chance and necessity." The aphorism that "necessity is the mother of invention" finds its counterpoint in Mark Twain's quip that "necessity is the mother of taking chances." Even in our most goal-directed endeavors, we see the tension between accident and purpose, as Louis Pasteur did in saying that "chance favors only the prepared mind."

My intention in writing this chapter is not to sort out the tangled nuances of the words *chance* and *necessity*. Nonetheless, it might be helpful to illustrate some of these nuances before proceeding. Chance often invokes some instantaneous disturbance, such as a cosmic ray striking a chromosome and causing a particular mutation. Chance is also sometimes used with reference to contingent effects of prior historical events, such as how the course of life on Earth might have unfolded differently had some asteroid not caused a certain mass extinction. Yet, the cosmic ray may have followed a path set by the laws of physics, and the historical influences might have been inevitable in their time. What the notion of chance captures is the sense of unpredictability and the absence of control exerted by the affected system over its own eventual fate.

Necessity is fraught with even more divergent meanings. Necessity is often used to describe outcomes that are inevitable given the action of physical laws, such as the motion of one billiard ball that has been struck by another ball at a particular angle and momentum. Necessity can also refer to a purposeful course of action, one that must be followed in order to achieve some desired end, such as striking one ball with a cue so that it hits another ball at the angle and momentum that is required to move the second ball in a particular way. And in an evolutionary context, necessity provides a shorthand term to describe adaptive solutions, produced by natural selection, that allow organisms to cope with the various challenges they face in their environments. The hand-eye coordination that enables the pool player to strike a ball precisely as intended might be an adaptation that was necessary for survival during some part of the history of our species. (Ayala [1999], Pennock [1999], and Ruse [2003] discuss important similarities and differences between designs produced by the deliberate actions of conscious agents and those that result from natural selection.)

The Roles of Chance and Necessity in Evolutionary Thought

The tension between chance and necessity is perhaps more central to evolutionary biology than to any other science. Physics certainly encompasses the determinism of classical mechanics and the randomness of quantum mechanics, but these forces play out at such different scales that the difficulty lies in linking these two realms rather than in disentangling their effects. By contrast, the tension between chance and necessity enters into current evolutionary thought at two levels that are both central to our understanding of the biological world in which we live.

At one level, we have the historical narrative of life on Earth that is the primary focus of paleontological and much comparative research. It is a great struggle, of course, to sort out what happened and when, especially across the vast reaches of time. Nonetheless, things really did happen and at certain moments. Thus, there is only one true history that occurred, although we will never be able to reconstruct it in its entirety. But just as students of human history are fond of asking how things might have unfolded differently if some past event were altered, so too evolutionary biologists are fascinated by similar questions. Stephen Jay Gould, in *Wonderful Life* (1989), offered the thought experiment of "replaying life's tape" to evoke these what-if questions in the context of evolution. What if different animal phyla had survived the Cambrian than those that did? What if an asteroid had not hit Earth at the end of the Cretaceous? Or what if the asteroid had been half the size, or twice the size, of the one that actually hit? For that matter, what if it had hit just one hour sooner or later? What difference would these accidental circumstances have made to the subsequent evolution of life, including our own coming into being? For Gould, the quirks of history and the immensity of alternative paths led him to infer the "awesome improbability of human evolution"—not only in the narrow sense of our particular species but more generally in the sense of any species that can wonder and reason about its own origins. Most evolutionists accept Gould's conclusion in the narrow sense, but others have argued against his more general conclusion. Simon Conway Morris (2003) presents myriad examples of parallel and convergent evolution, whereby multiple lineages have independently evolved similar adaptations to similar challenges (such as eyes to detect light). He then uses this repeatability to argue that any general features of organisms that are of great adaptive value (and that are genetically accessible) would have arisen, sooner or later, and human-like intelligence is unlikely to be an exception.

The second level of interplay between chance and necessity lies at the heart of the mechanistic basis of Darwinian evolution itself. Natural selection, of course, provides the directing force by which organisms acquire features that fit them to their environments. Those individuals that have certain phenotypic features are more successful in the struggle for survival and reproductive success than others that have different features. If the phenotypic differences are heritable, then those features that enhance performance will tend to be amplified in later generations, giving the appearance of direction, design, and purpose. Heritable differences between organisms are encoded in their genomes, and these differences are produced by recombination and mutation. Sexual recombination scrambles the existing differences between two parental genomes, while mutation provides the ultimate source of this genetic variation. It is at the level of mutation that Darwinian evolution is, in essence, random and accidental.

Let me be clear with respect to what evolutionary theory means—and does not mean—when we say that mutations are random and occur by chance. We do not mean that mutations occur at the same rate throughout a genome; in fact, some DNA sequences are more mutable than others. Nor do we mean that the environment plays no role in causing mutations; it does, as witnessed by mutagenic agents. Nor do we mean that organisms can exert no control whatsoever over the mutational process; in fact, organisms from bacteria to humans possess exquisite molecular machinery for proofreading their DNA and correcting errors during replication. What is important, however, is that mutations are random insofar as organisms cannot direct the production of particular mutations in response to their particular needs. (Humans, through the tools of genetic engineering, are on the cusp, for better or worse, of directing some of our own mutations.) Thus, mutations are genetic accidents, and they do not provide the design-like directionality given by natural selection. However, in their scatter-shot way, mutations provide the heritable variation that is needed for selection to proceed. Because more mutations are deleterious than are beneficial, much of natural selection consists of eliminating deleterious mutations. But some mutations produce useful features, and these have fueled the adaptation of organisms to their environments.

Charles Darwin is justifiably renowned for presenting a coherent body of evidence to support the general proposition of organic evolution and especially for discovering the principle of natural selection. But he was largely ignorant of hereditary mechanisms, including what we now call mutation. Even so, there was an important aspect of his reasoning that I think is not nearly as well recognized as it should be. That is, Darwin was remarkably clear in distinguishing between what he did understand—how natural selection could improve fitness across generations—and what he could not understand—the source of the variation on which selection acted. His chapter on "Laws of variation" (1859, 170) concludes as follows: "Whatever the cause may be of each slight difference in the offspring from their parents—and a cause for each must exist—it is the steady accumulation, through Natural Selection, of such differences, when beneficial to the individual, that give rise to all the more important modifications of structure, by which the innumerable beings on the face of this earth are enabled to struggle with each other, and the best adapted to survive." At the outset of this same chapter (131), Darwin describes variation as being "due to chance" but adds, "This, of course, is a wholly incorrect expression, but it serves to acknowledge plainly our ignorance of the cause of each particular variation." Thus, a key to Darwin's success was his ability to separate what he understood from what he did not. (George Zebrowski [2000], a science-fiction author, beautifully captured the fundamental strength and limitation of science in a maxim he attributed to the cosmologist Hermann Bondi: "The power of science comes from being able to say something, without having to say everything." Darwin was able to say something powerful and profound about the consequences of heritable variation, even while he humbly and forthrightly admitted his ignorance about the underlying causes of that variation.)

For several decades after the rediscovery of Mendel's findings on particulate inheritance, it was widely accepted that mutations were random events in the sense that I discussed above. Indeed, many mutations were demonstrably harmful to the organisms that carried them, and so it made little sense to think of them as somehow directed toward producing adaptation. However, it was difficult to test this assumption formally because most populations of experimental organisms, such as fruit flies, had substantial standing variation, thus making it almost impossible to distinguish new mutations from rare variants already present. Things were even more confused for those who worked with bacteria, where it was impossible to see individual mutants or demonstrate their existence except by imposing selection for some new phenotype. When such selection was imposed and the bacteria acquired a new phenotype, one could not tell if selection had caused the phenotypic conversion of the entire population or, alternatively, if selection had allowed some rare mutant type to take over the population. One microbiologist of that era expressed the discord as follows (Lewis 1934, 636): "The subject of bacterial variation and heredity has reached an almost hopeless state of confusion. Almost every possible view has been set forth, and there seems no reason to hope that any uniform consensus of opinion may be reached in the near future. There are many advocates of the Lamarckian mode of bacterial inheritance, while others hold to the view that it is essentially Darwinian." A while later, Julian Huxley (1942, 131–132) explicitly excluded bacteria from the then-modern evolutionary synthesis by saying, "They have no genes in the sense of accurately quantized portions of hereditary substance. . . ."

Ironically, just one year after Huxley excluded bacteria from the emerging evolutionary synthesis, the biologist Salvador Luria and the physicist-turned-biologist Max Delbrück published one of the great experiments of all time, which demonstrated that bacterial mutations do, in fact, occur at random (Luria and Delbrück 1943). Without going into the details of their subtle and elegant experiment, they showed that mutations that conferred resistance on bacteria to lethal infections by viruses had occurred in generations prior to the bacteria's exposure to the viruses; hence, the mutations could not have been caused by that exposure, and they must have arisen spontaneously without regard to their utility. Further experiments performed by Joshua Lederberg and Esther Lederberg (1952) supported the same conclusion, and they did so in a way that made a striking visual impression on anyone who remained skeptical of the quantitative reasoning necessary to interpret the experiment of Luria and Delbrück. With these experiments, mutation and selection were firmly established as the biological processes that correspond, respectively, to chance and necessity. (Again, by saying that mutations are due to chance, one does not imply that mutations lack physical causes. A certain mutation might have been caused by a cosmic ray hitting a particular site on a chromosome. But such physical events are beyond the control of the affected organism, in the same way that a gambler does not control the outcome of a throw of the dice, even though dice also obey ordinary physical laws.)

Putting the Powers of Chance and Necessity to the Test

So far in this chapter, I have touched on some important lines of biological thought on the ideas of chance and necessity and their evolutionary significance, ranging from experimental research focused on the origins of mutations to paleontological and comparative perspectives on the potential macroevolutionary consequences of chance and necessity. I will now summarize some of my group's research in this area, which attempts to bridge perspectives and time scales by bringing the macroevolutionary framework on contingency versus repeatability down to an experimental scale. These experiments allow us to watch phenotypic and genomic evolution across thousands of generations. Also, the experiments involve replicate populations that begin with the same ancestor and evolve in identical environments, such that we can characterize both parallel and divergent changes. And the system can be preserved at intermediate stages, enabling us to rewind and restart the evolutionary tape in order to place hypotheses that invoke historical contingency into the same framework as those that invoke adaptation. With these motivations, I will now discuss an experiment with the bacterium *Escherichia coli* that has been underway in my laboratory for almost two decades.

E. coli has a number of features that make it well suited for experiments to investigate evolutionary dynamics and outcomes. This species is easy to propagate and enumerate; one can control and manipulate environmental factors; its generations are rapid and population sizes are large; and it reproduces asexually by binary fission. Moreover, one can preserve and later revive ancestral genotypes as well as those from intermediate times in an experiment. This last feature, coupled with suitable genetic markers, allows us to measure the extent of adaptation by allowing derived genotypes to compete against their own ancestors. Several decades of intensive research on the physiology and genetics of *E. coli* provide a wealth of information on the inner workings of its cells, while various molecular biological tools permit precise genetic analysis and manipulation.

In the long-term experiment, twelve populations were founded from single cells of the same ancestral strain, and the populations have now been propagated for more than 40,000 bacterial generations in identical environments (Lenski 2004). The environment consists of a simple medium with glucose as the sole source of carbon and energy available to the cells. Every day, each population is diluted one hundred-fold into fresh medium, where it grows to several hundred million cells before depleting the glucose and awaiting the next transfer. Because each population began as a single haploid cell, there was no variation either within or between populations at the outset (except a neutral genetic marker embedded within the design of the experiment). Therefore, all of the variation required for adaptation and divergence had to arise *de novo* by mutation, so that this experiment encompasses the origin as well as the fate of genetic novelties.

Given the population size and knowledge of mutation rates, it is likely that each population has had more than a billion mutations appear, even after taking into account the bottleneck effect during the daily transfers. And given the fact that the genome of *E. coli* is about five million base-pairs, it then follows that almost all mutations have been tried many times over in each population. However, the fact that each one-step move has been tried repeatedly does not imply that most genotypes have existed, as only a tiny "corner" of the immense genotypic space is ever probed in such an experiment. Moreover, most mutations are lost to genetic drift or natural selection, and I estimate that only tens or hundreds of mutations have been substituted in a typical population (Lenski 2004).

Have the replicate populations adapted in similar or different ways? In other words, how have chance and necessity played out in this simple little world? To answer these questions, we have sought to characterize both phenotypic and genomic changes, and I will now highlight some of the main findings published to date, as well as some recent findings not yet published. All twelve populations have improved substantially in fitness, such that after 20,000 generations they grew, on average, about 70 percent faster than their common ancestor in direct competitions (Cooper and Lenski 2000). Moreover, all twelve populations had similar fitness trajectories, with the rate of improvement much greater early in the experiment and decelerating as it continued (Lenski and Travisano 1994; Cooper and Lenski 2000). All twelve derived populations also produce cells that are much larger than the ancestral cells, although the between-lineage variation in size and shape is greater than the variation in their competitive fitness (Lenski and Travisano 1994; Lenski and Mongold 2000). Also, all twelve populations have tended to become glucose specialists, insofar as their performance capacities on a diverse array of other substrates tended to narrow as they adapted to glucose (Cooper and Lenski 2000). However, the details of their correlated changes in performance on other substrates have varied considerably across populations. For example, when competitions between evolved and ancestral genotypes were performed on either glucose or maltose, the derived populations were far more variable in their relative fitness levels on maltose than on glucose (Travisano and Lenski 1996). This particular specificity of adaptation is interesting given that maltose is, in fact, di-glucose. We examined the genome-wide profiles of gene expression in the ancestor and two of the evolved lineages at the mRNA and protein levels (Cooper et al. 2003; Pelosi et al. 2006). These profiles showed strikingly parallel changes at both levels, such that the two populations that had evolved independently for 20,000 generations from a common ancestor were much more similar to one another in their overall expression patterns than they were to their ancestor.

Summarizing thus far, the phenotypic data have tended to emphasize the power of "necessity" to produce parallel changes across independent lineages that experienced the same selective regime. However, some "chance" differences, often subtle, have also emerged between these lineages.

Our next objective has been to extend our analyses of these populations to the molecular-genetic level, in order to see if the extensive phenotypic parallelism also extends to the genes. Have the same mutations occurred and been selected in the independently evolving lineages? Or perhaps have different mutations evolved but in the same genes and pathways? Or does the phenotypic parallelism mask bewilderingly idiosyncratic responses at the genetic level? I should begin by emphasizing that it is difficult to find mutations in these populations; as noted earlier, the *E. coli* genome contains millions of base-pairs, and only a few tens or hundreds of all the mutations that occurred have been substituted in a population. One approach we have pursued, which serves as a useful control, was to choose thirty-six gene regions at random and sequence those regions in clones sampled from all twelve populations. This approach yielded only a few mutations, and in no case did we find the same gene bearing a mutation in even two of the twelve lineages (Lenski et al. 2003). Thus, the background rate of genomic change was indeed low. By contrast, when we have used parallel phenotypic changes, such as changes in resource usage and gene expression, to suggest candidate genes for study, we have found genes that underwent changes in many or even all of the lineages. For example, all twelve populations have deletions affecting the ribose operon (Cooper et al. 2001); and eight have point substitutions in *spoT*, which encodes a global regulator of gene expression (Cooper et al. 2003). For both these cases, we moved one of the evolved alleles into the ancestor to confirm that the substituted mutation is indeed beneficial in the environment where it evolved. Interestingly, when we moved one of the evolved *spoT* alleles into another evolved lineage that had retained the ancestral *spoT* allele, the evolved allele did not confer any advantage. This other lineage evolved similar changes in its global expression profile, implying that a mutation in some other (as yet unidentified) gene must produce similar effects on both gene expression and fitness, so that the *spoT* mutation was rendered redundant and, therefore, not beneficial. Several colleagues and I have now identified several more genes in which many or all of the evolved populations have substituted mutations (Crozat et al. 2005; Pelosi et al. 2006; Woods et al. 2006). Although the same genes changed in multiple lineages, the precise mutations that were substituted are different, with only rare exceptions, at the nucleotide level. Summarizing these genetic data, the overall conclusion seems to be reasonably concordant with that based on the phenotypic patterns. That is, the adaptive substitutions are concentrated in a few genes, emphasizing again the power of selective "necessity" to produce parallel changes even at the genetic level. At the same time, the particular alleles that arose and the exact subset of genes that changed in any given lineage indicate the importance of "chance" mutations in promoting evolutionary divergence even under identical selective conditions.

There is one striking exception, however, to the overall pattern of repeatable and parallel evolution that I summarized above—an exception that puts far more emphasis on the view that important evolutionary transitions may depend on the accidents of history. This exception is also an area of ongoing (and not yet published) work in my laboratory; therefore, I will avoid presenting some details while giving an overview of the situation and how we intend to explore further the evolutionary interpretation and significance of this latest finding. At the outset of describing this experiment, I mentioned that glucose provided the sole source of carbon and energy available to the cells. However, citrate was also present in the medium throughout the experiment but was unavailable to the cells because they could not use it; in fact, the inability to grow on citrate in an aerobic environment is an important diagnostic feature of *E. coli,* whereas many related species can use citrate. For many years, I wondered if the evolving populations would discover some way to use this second resource, but none did so for more than 30,000 generations. This inability persisted even though each population should have tested almost every possible one-step mutation many times over, the resulting benefit of using this unexploited resource would be large, and the rate of continued adaptation to other aspects of the experimental environment had slowed down substantially.

Then, around 32,500 generations, one of the twelve populations evolved the ability to use the citrate in the medium. At first, I suspected some citrate-utilizing bacterium had contaminated one of the populations, but this possibility has been excluded; the citrate-utilizing form is, in fact, a derivative of the particular *E. coli* strain used in this experiment. My student Zachary Blount and I have formulated two distinct hypotheses for why this major transition occurred in only one population and only after so much elapsed time. We are eager to determine which hypothesis is correct because the difference between them cuts sharply across the question of repeatability versus contingency in evolution.

According to one hypothesis, the mutation that produced the citrate-using phenotype is extremely rare, much more so than implied by the calculation that almost all one-step mutations should have occurred many times over in each population. Perhaps, for example, the mutation is some inversion in which both end points of the inverted region must occur at precise locations. Under the second hypothesis, the mutation that ultimately yielded the citrate-using phenotype was not rare or unusual in and of itself, but it interacted in some particular way with one of the previously evolved idiosyncratic differences between this lineage and the others, despite the mostly parallel nature of their preceding evolution. Thus, the first hypothesis invokes a very rare, but ultimately repeatable, change. The second hypothesis, by contrast, invokes a contingent series of changes, such that the potential for evolution to yield profoundly divergent outcomes depends on subtle and—at the time, inconsequential—differences in the precise steps taken along nearly parallel pathways.

We are now pursuing two experimental strategies to test these hypotheses. One strategy aims to find the mutation that was proximally responsible for producing the first citrate-using cells. We could then move that mutation into the other evolved genomes and ask whether it also converts them to citrate users. If it does, then the question becomes: why is the mutation so rare that it had not already occurred and been selected in the other lineages? But if transferring the mutation that conferred the citrate-using ability on one lineage does not confer that ability on the other lineages, then this result would imply that the citrate-using phenotype required some prior genetic substitution that was unique to that one lineage.

The second strategy is, quite literally, to "rewind the tape of life" in this experiment and then restart many new populations from different intermediate time points in the lineage that evolved the citrate-using phenotype. If the evolution of that phenotype did not depend on an unusually rare mutation, but instead was conditional on some particular prior change, then we should find that the percentage of restarted populations that achieve the citrate-using phenotype changes over time. For example, imagine that we restart six populations each from clones sampled at generations 30,000 and 32,000. Now imagine that none of the populations restarted from generation 30,000 evolves the citrate-using phenotype during the next 3,000 generations, while all six populations restarted from generation 32,000 evolve that phenotype over the same time. This outcome would strongly indicate that some mutation was substituted between 30,000 and 32,000 generations in the lineage that later evolved the citrate-using phenotype—a pivotal mutation that did not yield this novel phenotype but which predisposed its subsequent evolution. In principle, we might be able to find this predisposing mutation, move it into the genomes of the other lineages, and show that they, too, could then evolve the citrate-using phenotype. But we might also discover that the predisposing mutation was not beneficial in the other lineages, perhaps because other mutations conferred the same immediate benefit without predisposing the subsequent evolution of the citrate-using phenotype. In that case, we would have to conclude that the largely parallel and repeatable evolutionary changes had nonetheless pushed the replicate populations onto divergent paths leading to very different final outcomes.

Perspectives

The paleontologist George Gaylord Simpson (1944, xvii) said that experiments "may reveal what happens to a hundred rats in the course of ten years under fixed and simple conditions, but not what happened to a billion rats in the course of ten million years under the fluctuating conditions of Earth history." Although I much prefer my evolving bacteria to rats, I accept the difficulty, even impossibility, of extrapolating from any experiment to the grand sweep of life's history on Earth. Nonetheless, experiments encourage clarity about particular expectations under alternative hypotheses. In this context, even claims about subsequent consequences of historical events can be subjected—by rewinding and replaying life's tape—to the same scientific criteria of prediction and hypothesis-testing as are claims of adaptation by natural selection. Finally, the surprisingly rich and complex interplay between chance and necessity that has emerged in this one little experiment gives me even greater respect for the essential contributions of both chance and necessity to the history of life on Earth, including our own coming into being.

Acknowledgments

I thank Simon Conway Morris for inviting me to participate in the symposium that led to this book, Mary Ann Meyers for skillfully organizing that symposium, and George Coyne for his hospitality during the symposium at Castel Gandolfo. I also thank Shoshannah Lenski and Rob Pennock for providing helpful comments on this chapter and Zachary Blount and Chris Borland for allowing me to discuss some results from our work that is in progress. My research has been supported primarily by the U.S. National Science Foundation.

References

Ayala, F. J. 1999. Adaptation and novelty: Teleological explanations in evolutionary biology. *History and Philosophy of the Life Sciences* 21: 3–33.

Conway Morris, S. 2003. *Life's solution: Inevitable humans in a lonely universe.* Cambridge: Cambridge University Press.

Cooper, T. F., D. E. Rozen, and R. E. Lenski. 2003. Parallel changes in gene expression after 20,000 generations of evolution in E. coli. *Proceedings of the National Academy of Sciences, U.S.A.* 100: 1072–77.

Cooper, V. S., and R. E. Lenski. 2000. The population genetics of ecological specialization in evolving *E. coli* populations. *Nature* 407: 736–39.

Cooper, V. S., D. Schneider, M. Blot, and R. E. Lenski. 2001. Mechanisms causing rapid and parallel losses of ribose catabolism in evolving populations of *E. coli* B. *Journal of Bacteriology* 183: 2834–41.

Crozat, E., N. Philippe, R. E. Lenski, J. Geiselmann, and D. Schneider. 2005. Long-term experimental evolution in *Escherichia coli*. XII. DNA topology as a key target of selection. *Genetics* 169: 523–32.

Darwin, C. 1859. *On the origin of species*. London: John Murray.

Gould, S. J. 1989. *Wonderful life: The Burgess Shale and the nature of history*. New York: Norton.

Huxley, J. 1942. *Evolution: The modern synthesis*. New York: Harper.

Lederberg, J., and E. M. Lederberg. 1952. Replica plating and indirect selection of bacterial mutants. *Journal of Bacteriology* 63: 399–406.

Lenski, R. E. 2004. Phenotypic and genomic evolution during a 20,000-generation experiment with the bacterium *Escherichia coli*. *Plant Breeding Reviews* 24: 225–65.

Lenski, R. E., and J. A. Mongold. 2000. Cell size, shape, and fitness in evolving populations of bacteria. In *Scaling in biology*, ed. J. H. Brown and G. B. West, 221–35. Oxford: Oxford University Press.

Lenski, R. E., and M. Travisano. 1994. Dynamics of adaptation and diversification: A 10,000-generation experiment with bacterial populations. *Proceedings of the National Academy of Sciences, U.S.A.* 91: 6808–14.

Lenski, R. E., C. L. Winkworth, and M. A. Riley. 2003. Rates of DNA sequence evolution in experimental populations of *Escherichia coli* during 20,000 generations. *Journal of Molecular Evolution* 56: 498–508.

Lewis, I. M. 1934. Bacterial variation with special reference to behavior of some mutable strains of colon bacteria in synthetic media. *Journal of Bacteriology* 28: 619–38.

Luria, S. E., and M. Delbrück. 1943. Mutations of bacteria from virus sensitivity to virus resistance. *Genetics* 28: 491–511.

Pelosi, L., L. Kühn, D. Guetta, J. Garin, J. Geiselmann, R. E. Lenski, and D. Schneider. 2006. Parallel changes in global protein profiles during long-term experimental evolution in *Escherichia coli*. *Genetics* 173: 1851–69.

Pennock, R. T. 1999. *Tower of Babel: The evidence against the new creationism*. Cambridge, MA: MIT Press.

Ruse, M. 2003. *Darwin and design: Does evolution have a purpose?* Cambridge, MA: Harvard University Press.

Simpson, G. G. 1944. *Tempo and mode in evolution*. New York: Columbia University Press.

Travisano, M., and R. E. Lenski. 1996. Long-term experimental evolution in *Escherichia coli*. IV. Targets of selection and the specificity of adaptation. *Genetics* 143: 15–26.

Woods, R., D. Schneider, C. L. Winkworth, M. A. Riley, and R. E. Lenski. 2006. Tests of parallel molecular evolution in a long-term experiment with *Escherichia coli*. *Proceedings of the National Academy of Sciences, USA* 103: 9107–12.

Zebrowski, G. 2000. The holdouts. *Nature* 408: 775.

2 CONVERGENT EVOLUTION

A Periodic Table of Life?

George McGhee

"Convergent Evolution" and Predictability in Chemistry

Are there deeper laws to biological organization? I suspect that the modern scientific discipline of evolutionary biology is in a similar position as the scientific discipline of chemistry before the discovery of the periodic table of elements. The periodic law was discovered by Dmitri Mendeleev in 1869 (if I remember correctly, Mendeleev maintained that the idea came to him in his sleep!). Before this time, the complexity of chemical reactions made little sense. It was known that some elements were very reactive and combined readily. On the other hand, other elements were inert or rarely combined with other elements. Each element seemingly had its own, unique behavior—much as it seems today that each species of life has its own individual nature, the product of its unique evolutionary history.

Mendeleev discovered that a simpler, deeper structure underlay the apparent unique complexity of the elements. The chemical properties, or behaviors, of the elements recur periodically when the elements are simply arranged in the increasing order of their atomic numbers, of the number of protons in the nucleus of the elemental atom. For example, helium, neon, argon, krypton, and xenon are all unique elements. Yet their chemical behavior is very similar, and we unite them together as "noble gases" in the periodic table of elements.

The evolution of the universe began with the element hydrogen, with one proton. The element helium evolved via the process of fusing two hydrogen atoms together in the first generation of stars in the universe, producing a new atom with two protons in the nucleus. This same evolutionary process eventually produced the element neon, with ten protons; argon, with eighteen protons; krypton, with thirty-six protons; and xenon, with fifty-four protons. Each element has its own unique evolutionary history: some were formed in the first-generation stars, while others formed in second-generation stars, formed from the debris of the explosion of first-generation stars. Although each of these elements is unique, with its own evolutionary history, their chemical behaviors can be viewed as *convergent*—they are all inert gases.

Mendeleev was the first to discover that the convergent behavior of these gases was a function of their atomic number; that is, gases with two, ten, eighteen, thirty-six, and fifty-four protons in their atomic nuclei all behaved similarly, even though each element is an individual with its own unique evolutionary history. Arranging the elements in the periodic table revealed other groups of elements that had evolved convergent behaviors, such as carbon and silicon, and so on. Suddenly, the bewildering complexity of elemental chemistry was revealed to have a *simple underlying structure*, allowing us to *predict their behavior* based upon their position in the table.

Convergent Evolution in Biology and Predictability in Evolution

The evolution of living organisms is much more complex that the evolution of the elements of the universe. But does a simpler structure underlie the bewildering complexity of organisms, similar to the simpler structure that underlies the complexity of the elements? I suggest that the phenomenon of convergent evolution hints that this might be the case.

Predictability in evolution is the key concept that is linked to the phenomenon of convergent evolution. Most natural selection theoreticians routinely state that biological evolution is unpredictable. For example, see the otherwise delightful cartoon booklet by Jay Hosler (2003), which attempts to explain the theory of natural selection to a general, nonspecialist audience. In it, the author has chosen a highly unusual medium to convey his ideas, in that the cartoon character of Darwin conducts discussions about the implications of his theory with hair-follicle mites that live in his eyebrows—mites that, moreover, believe Darwin to be God! In one scene in the book, Darwin patiently explains to the mites, who consider evolution to be progressive, that "evolution is not a nice, neat progressive march. There's no predictable destination" (Hosler 2003). While I firmly agree with the former sentence, I also equally firmly disagree with the latter.

In contrast, I argue that the oft-repeated statement that "biological evolution has no predictable destination" is demonstrably false. Consider one of the most frequently cited cases of convergent evolution: the astonishing morphological similarity between the extinct Mesozoic marine reptile *Ichthyosaurus* and the living marine mammal *Phocaena*, the porpoise, and *Delphinus*, the dolphin. Not only do they look amazingly similar to one another, but they all look amazingly similar to large, fast-swimming fish like the tuna or swordfish. The cartilaginous fish and the bony fish both solved the physics of swimming back in the Silurian by evolving streamlined, fusiform morphologies (Figure 1). Some 230 million years later, a group of land-dwelling reptiles rediscovered this same morphology in their evolutionary return to the sea (Figure 1). And around 175 million years later, a group of land-dwelling mammals also rediscovered this same morphology in their own evolutionary return to the sea (Figure 1).

The evolution of an ichthyosaur or porpoise morphology is not trivial. It can be correctly described as nothing less than astonishing that a group of land-dwelling tetrapods, complete with four legs and a tail, could devolve their appendages and their tails back into fins like those of a fish. Highly unlikely, if not impossible? Yet it happened *twice*, convergently in the reptiles and the mammals, two groups of animals that are not closely related. We have to go back in time as far as the Carboniferous to find a common ancestor for the mammals and the reptiles; thus, our genetic legacies are very, very different. Nonetheless, the ichthyosaur and the porpoise both have independently reevolved fins.

Contrary to the dictum that "biological evolution has no predictable destination," I predict with absolute confidence that if any large, fast-

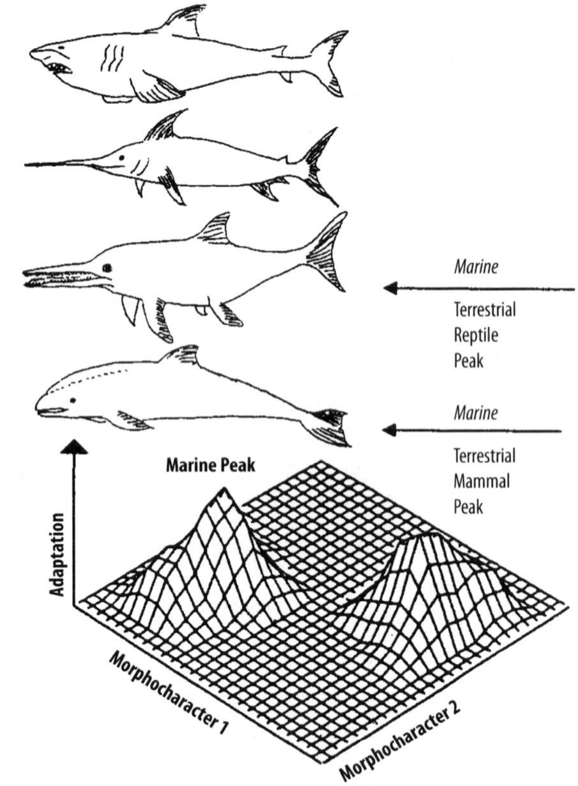

Figure 1. An adaptive landscape representation of the convergent evolution of streamlined, fusiform swimming morphologies in the cartilaginous fish (a shark, top left in the figure), in the bony fish (a swordfish, second down on the left), in the reptiles (an ichthyosaur, third down on the left), and in the mammals (a porpoise, fourth down on the left). In the adaptive landscape model, topographic highs (peaks) represent the coordinates of morphologies with optimal function, slopes on the topographic hill represent morphologies with less-than-optimum function, and coordinates in the flat plain represent nonfunctional morphologies. Modified from McGhee (2007).

swimming organisms exist in the oceans of the moon Europa—far away in orbit around Jupiter, swimming under the perpetual ice that covers their world—then they will have streamlined, fusiform bodies; that is, they will look very similar to a porpoise, an ichthyosaur, a swordfish, or a shark (Figure 1).

Rerun the Tape of Life?

The best-known evolutionary essayist of the twentieth century, the Harvard paleontologist Stephen Jay Gould, was fond of a thought experiment of his own that he called "replaying life's tape" (Gould 1989). That is, consider the history of the evolution of life on Earth to be similar to a video tape of a popular movie. Then, imagine what would happen if you could take a copy of the video tape and rewind it to a point early in the movie, erasing everything on the tape that happened after that point, and rerun the tape to see what happens a second time. Will the historical sequence of events in the evolution of life in the second rerun of the tape resemble the original? Or will evolution take radically different pathways in the second rerun, producing animal and plant forms totally unlike those of the original? Gould argued strongly for the second scenario: "Any replay of the tape would lead evolution down a pathway radically different from the road actually taken. . . . The diversity of possible itineraries does demonstrate that eventual results cannot be predicted at the outset. Each step proceeds for cause, but no finale can be specified at the start, and none would ever occur a second time in the same way, because any pathway proceeds through thousands of improbable stages" (Gould 1989, 51).

If I were to insist to a chemist that if he or she were to rerun the process of the evolution of the universe—go all the way back to the big bang and start all over again—that the elemental composition of the universe would be entirely different, that it is highly unlikely that neon or argon would be present in that new universe, I am certain I would be promptly escorted out of his or her laboratory as a person with obviously no knowledge of science. Chemists know that, if you start again with an atom with one proton, hydrogen, that the process of stellar atomic fusion will eventually produce an atom with two protons, helium, and that eventually neon and argon would reevolve. It is not an absolute certainty—one could imagine a new universe that is a uniformly distributed cloud of hydrogen gas that never collapses into star formation—yet it is highly likely that stars will form, and then it is a virtual certainty that argon would reappear.

Is the evolution of life so different? Is evolution such a chance phenomenon, such a random series of unconstrained events, that, if we reran the process of biological evolution, the organisms present in that new universe would be entirely different? Biological evolution is vastly more complex than elemental evolution, but is not the process similar, at least in the initial stages? Elemental fusion is a process that is at least conceptually similar to organic symbiosis. If we started again with a single simple prokaryote cell, a bacterium, would not the process of symbiosis again produce a more complex eukaryote cell, just as the fusion of two hydrogen atoms will again produce an atom of helium? Take the initial bacterial cell and add, via symbiosis, a cyanobacterial cell to it, and you once again have a more complex cell with a chloroplast. Add a purple bacterium, and you have a more complex cell with both chloroplast and mitochondrion. Eventually fuse these eukaryote cells together, and you reevolve a multicellular form of life.

If stars form in our new universe, I would predict that neon and argon would reevolve. If life forms in our new universe, I would predict that multicellularity would reevolve.

A Periodic Table of Life?

The ancient Greeks first conceived of the concept of the atom (it is usually attributed to Democritus, ca 460–ca 352 BCE). It took over two thousand years before Mendeleev came up with the concept of arranging those atoms into the periodic table of elements. If a "periodic table of life" exists, hopefully it will not take as long to discover it as it took to discover the periodic table of elements.

The noble gases all behave in a similar convergent fashion due to a similar distribution of the electric fields of their atoms. The shark, swordfish, ichthyosaur, and porpoise all behave in a similar fashion due to the similar demands of the physics of swimming. The periodic table of elements was discovered by analyzing the convergent behavior of the elements—I hope that we can discover a simpler structure underlying the complexity of evolution by analyzing the convergent behavior of species.

In a simple thought experiment, it is easy to construct a preliminary "periodic table of animals" (Table 1). In essence, the various rows of the periodic table of elements are based upon the "complexity" of the atomic structure of the elements: elements in the first row have only the electron-shell K; elements in the second row have the electron-shells K and L; third-row elements have electron shells K, L, and M; and so on. These rows also reflect the evolutionary sequence of appearance of the elements, with elements in the first row (hydrogen and helium) appearing first in the evolution of the universe, elements in the second row evolving next,

Table 1. A "periodic table of animals," based upon locomotory type and evolutionary sequence of origination.

SPECTRUM OF LOCOMOTION

SEQUENCE OF EVOLUTION:	2-D Locomotion Crawling (Legless)	Walking (Legs)	3-D Locomotion Swimming (Fusiform body)	Flying (Wings)
1. Invertebrates	Annelids, Gastropods	Arthropods	Cephalopods	Insects
2. Amphibians	Caecilians	Amphibians	Tadpoles	—
3. Reptiles	Snakes	(reptiles)	Ichthyosaurs, Mosasaurs	Pterosaurs
4. Dinosaurs	—	(dinosaurs)	—	Birds
5. Mammals	—	(mammals)	Porpoises, Whales	Bats

and so on. We can use these elemental concepts of "complexity and evolutionary sequence" in an analogous fashion by arranging the major groups of animals in a similar series of rows (Table 1).

The various columns of the periodic table of elements can be considered to characterize the "mobility" of the elements in those rows, with highly mobile elements in some columns (elements that chemically combine readily, such as the column containing hydrogen, lithium, sodium, etc.) and elements that have low mobilities in other columns (elements that are chemically inert, such as the column containing helium, neon, argon, etc.). In an analogous fashion, we can consider the "mobility" of major animal groups on the basis of locomotory type in a series of columns (Table 1).

Even such a simple attempt to create a periodic table of animals immediately reveals major incidences of convergent evolution (Table 1). The previously discussed example of convergent evolution of fast-swimming fusiform morphologies in vertebrates (reptilian ichthyosaurs and mammalian porpoises) not only is apparent, but also we see that certain invertebrate animals have also convergently evolved this same fast-swimming morphology (most notably in modern-day squid and cuttlefish cephalopods and their extinct orthoconic and belemnitellid relatives). We see major convergences in the evolution of wing structures for powered flight: three separate and independent modifications of vertebrate forelimbs to wing structures in reptiles (pterosaurs), dinosaurs (birds), and mammals (bats), and the independent convergent evolution of similar invertebrate wing structures in flying insects. Does this not show us a predictable destination in evolution? If life evolves organisms capable of powered flight on an Earth-like world (similar gravitational field and atmosphere density) elsewhere in the universe, can we not predict that those organisms will have wings like those found on a bird, bat, or butterfly?

Two major groups of animals have convergently evolved leg structures for walking: the arthropods and the ancestral amphibians (the tetrapod reptiles, dinosaurs, and mammals are listed in parentheses in the walking column in Table 1 because their legs are not independent convergences but rather plesiomorphic structures simply inherited from their amphibian ancestors). Note, however, that both the amphibians and the reptiles have separately, convergently reevolved legless morphologies (amphibian caecilians and reptilian snakes) and morphologically resemble annelid worms!

One last point may be made with the simple periodic table of animals given in Table 1—that is, we can see predictable morphologies *that do not occur in nature*. We can predict what a legless, crawling dinosaur or mammal would look like, but such an animal has never evolved (or at least I am not aware of one). Some mammals are headed in this direction—weasels and ferrets come to mind, with their elongate bodies and small legs—but as yet a furry mammalian-snake form has yet to appear in the evolution of life on Earth. Likewise, the dinosaurs never returned to the sea (perhaps because successful, competitive marine reptiles already existed there), and the amphibians have never evolved powered flight (probably due to dehydration problems—there do exist, however, both frogs and snakes that glide). Yet we can easily *predict* what such a creature would have to look like, from a feathered snake to a flying frog. The ability to predict nonexistent biological form is one of the key features of the analytical techniques of *theoretical morphology*, techniques that might one day give us a periodic table of life.

A Research Program: Exploring the Spectrum of Existent, Nonexistent, and Impossible Biological Form

The rigorous analysis of convergent evolution requires us to try to visualize the theoretically possible pathways available to evolution: not only those evolutionary pathways that have led to convergent morphologies but also those pathways that have not been taken by evolution (McGhee 1999, 2001, 2007). In the evolution of organic form, how close is the match between the actual and the theoretically possible? What are the developmental pathways utilized by actual organisms through a hyperdimensional space of potential form, and what theoretically possible pathways are not found in existent organisms? What are the boundaries between possible and impossible biological form: can we reveal what biological forms nature *could produce* (regardless of what nature has actually produced), and can we reveal what biological forms *cannot exist at all*?

If we could do this, we would be on the pathway towards a periodic table of life. From the periodic table of elements, a chemist can tell you that, if we reran the evolution of the universe, the molecular "form" hydrogen hydroxide, HOH or H_2O, is chemically possible in our new universe. He or she could even tell you how the molecule would behave, that it would be polar, etc. On the other hand, the chemist could also tell you that the molecular "form" neon hydroxide, NeOH, is impossible in our new universe. Can we ever do the same for biological form? Can we predict what biological forms will occur in our new universe, and what will not?

I think we can: the analytical techniques of theoretical morphology provide a conceptual and computational basis to tackle such questions (Raup and Michelson 1965, McGhee 1999). Theoretical morphology involves the simulation of organic form by geometric or other mathematical models, producing either theoretical morphogenesis (hypothetical growth models) or theoretical morphospaces (hypothetical form distribution). Theoretical morphospaces are multidimensional spaces produced by systematically varying the parameter values of a mathematical model of form and are specifically produced without any measurement data from real organic form. As such, theoretical morphospaces are not only independent of existent morphology, but they can be used to create nonexistent morphology and also to identify regions of morphospace that contain geometrically impossible biological forms. One of the successes of early theoretical morphogenetic modeling was the demonstration that seemingly complex organic forms could be produced by relatively simple mathematical models, hinting that the actual developmental coding system may not be more complicated than the coding complexity of the computer simulation (Raup 1968).

What is the significance of the spatial distribution and density of forms within a theoretical morphospace? Specific to convergent evolution, what regions of the morphospace have been repeatedly, independently, occupied by organisms that started out from vastly different initial positions in the morphospace? The differential occupation of morphospace itself is neutral with respect to adaptation, and the theoretical morphospace concept differs from the concept of a fitness landscape, or adaptive landscape, in this regard (McGhee 1999, 2007). What are the evolutionary implications of empty morphospace—morphological pathways that are possible but have never been taken by evolution?

These questions are not only interesting from a biological perspective, but they have interesting philosophical implications as well (Maclaurin 2003). The important point here is that the analytical techniques of theoretical morphology allow us actually to ask, and answer, these questions for many organisms.

An Actual Example of Convergent Evolution in a Theoretical Morphospace

The geometry of the helix is ubiquitous in nature: an incredible number of biological structures on all scales, from molecules to entire animals, have evolved helical structures (the year 2003 was the fiftieth anniversary of the discovery that the coding mechanism of life itself, DNA, has a helical structure). Within the Bryozoa, a group of colonial marine organisms, helical colonies have convergently, independently, evolved in no less than six separate genera in distantly related higher taxa, scattered across a span of time comprising some 400 million years (McGhee and McKinney 2000; McKinney and McGhee 2003).

A two-dimensional slice through a three-dimensional theoretical morphospace of helical colony form in the Bryozoa is illustrated in Figures 2 and 3. Arranged around the two-dimensional morphospace axes are

Figure 2. A theoretical morphospace showing the boundary data polygons of 208 actual helical colony forms that have been evolved in seven different groups of marine bryozoans (from Raup, McGhee, and McKinney 2006). Computer simulations of existent colony form within the morphospace are also illustrated: the simulation in the upper right illustrates the morphology most frequently attained by convergent evolution, shown by the overlapping boundary polygons of morphologies evolved in four separate groups of bryozoans. The two simulations on the left and one simulation in the lower right illustrate more rarely evolved bryozoan morphologies (each present in only one species).

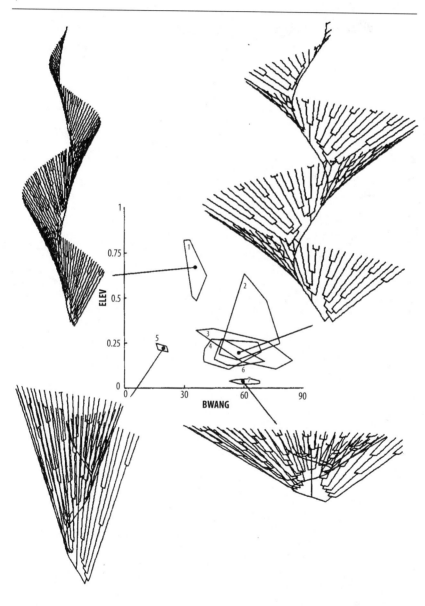

computer-simulated, hypothetical helical bryozoan colonies that have been produced with a mathematical model of helical geometry, and the position of each of the simulations within the morphospace is indicated (that is, which parameter value combinations will produce that simulation).

In Figure 2 are given the boundary polygons of measurement data taken from seven different groups of actual helical bryozoans, both extinct and alive, within the morphospace. Note the overlap region of the polygons in the center of the figure: these bryozoans have not only convergently evolved helical colonies, but they have *repeatedly evolved helical colonies that have the same geometry*, over and over again. The computer simulation given in the upper right of the figure illustrates this iteratively evolved geometry.

Note now the four computer simulations given in Figure 3. These simulations represent *nonexistent* colony morphologies; these four regions of the morphospace are empty of bryozoans. Thus, the theoretical morphospace can show us not only what organic form nature has produced over and over again, but it can also reveal to us a biological form that is theoretically possible but never produced by nature. Analysis of these nonexistent colony morphologies reveals that they represent nonfunctional geometries for the filter-feeding mode of life of marine bryozoans and that the iteratively reevolved colony form shown in Figure 2 is a product of functional constraint in bryozoan evolution (McGhee and McKinney 2000; McKinney and McGhee 2003).

The analytical techniques of theoretical morphology allow us to take the heuristic concept of evolution on an adaptive landscape (Figure 1) and to apply it to the analysis of the evolution of life (Figures 2 and 3). In essence, the overlapping boundary polygons of morphologies evolved within the Bryozoa in the past 400 million years, illustrated in Figure 2, are the apex and upper slope regions of an adaptive peak of helical colony form. The three colony morphologies shown on the left margin and lower right in Figure 2 represent the lower slope regions of the adaptive peak: they function, but not as well as the peak morphology, and each is only found in one species of bryozoan, respectively. And last, the four computer simulations given in Figure 3 show us the colony geometries that lie out on the flat plane of the adaptive landscape, the region of nonfunctional helical colony forms.

Figure 3. Computer simulation of nonexistent helical colony forms within the theoretical morphospace. Although these morphologies are geometrically possible, they have never evolved as organic forms within the bryozoans. From Raup, McGhee, and McKinney 2006.

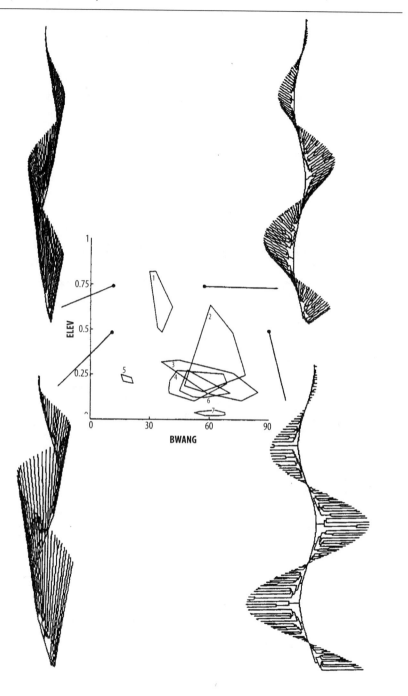

Conclusion: The Concept of Constraint in Evolution

Evolutionary constraint is a key phenomenon that underlies much of the convergent evolution that we see in nature and is, thus, a key to understanding the potential predictability of evolution and a possible periodic table of life. There are two general classes of evolutionary constraint: *extrinsic constraints*, those constraints imposed by the laws of physics and geometry, and *intrinsic constraints*, those constraints imposed by the biology of a specific organism. Extrinsic constraints exist whether any actual biological form encounters them or not, whereas intrinsic constraints do not exist in the absence of actual organisms. Two readily apparent extrinsic constraints in the evolution of life are *geometric constraints* and *functional constraints*. Likewise, at least two conceptually different types of intrinsic constraint exist: *phylogenetic constraint* and *developmental constraint* (McGhee 1999, 2007).

Theoretical morphospaces are particularly useful in exploring the limits of geometric constraint associated with a given morphogenetic model and in exploring the logical consequences of the model's fundamental assumptions. When we plot the actual distribution of the existent morphologies found in a group of organisms within the morphospace, we then may discover empty, unutilized regions of morphospace (Figure 3). As the empty regions of morphospace contain geometrically possible morphologies, the absence of these morphologies in existent organisms is not due to geometric constraint but may be due to functional constraint: it is possible that the morphologies found in the empty region of morphospace, while geometrically possible, are nonfunctional or of low fitness value. Alternatively, the observed nonexistent morphologies may be potentially functional but unattainable by a group of organisms due to their own intrinsic constraints: it is possible that the phylogenetic legacies, or *Baupläne*, of the organisms make the development of the geometries found in the empty region of the morphospace impossible.

For a given group of organisms, can we conceptually map the distribution and boundaries of developmental, phylogenetic, functional, and geometric constraints within theoretical morphospace? If we could accomplish this, we would be well on the way to understanding the reason that certain morphological solutions are repeatedly evolved in life, convergent evolution, as a function of the (I suspect) vastly larger areas of morphospace into which life cannot venture.

Acknowledgments

Figures 2 and 3 were originally published in the journal *Palaeontologia Electronica*.

References

Gould, S. J. 1989. *Wonderful life: The Burgess Shale and the nature of history*. New York: W. W. Norton & Company.

Hosler, J. 2003. *The sandwalk adventures: An adventure in evolution told in five chapters*. Columbus, OH: Active Synapse.

McGhee, G. R. 1999. *Theoretical morphology: The concept and its applications*. New York: Columbia University Press.

———. 2001. Exploring the spectrum of existent, nonexistent and impossible biological form. *Trends in Ecology and Evolution* 16: 172–73.

———. 2007. *The geometry of evolution: Adaptive landscapes and theoretical morphospaces*. Cambridge: Cambridge University Press.

McGhee, G. R., and F. K. McKinney. 2000. A theoretical morphologic analysis of convergently evolved erect helical colony form in the Bryozoa. *Paleobiology* 26: 556–77.

McKinney, F. K., and G. R. McGhee. 2003. Evolution of erect helical colony form in the Bryozoa: Phylogenetic, functional, and ecological factors. *Biological Journal of the Linnean Society* 80: 235–60.

Maclaurin, J. 2003. The good, the bad and the impossible. *Biology and Philosophy* 18: 463–76.

Raup, D. M. 1968. Theoretical morphology of echinoid growth. *Journal of Paleontology* 42: 50–63.

Raup, D. M., and A. Michelson. 1965. Theoretical morphology of the coiled shell. *Science* 147: 1294–95.

Raup, D. M., G. R. McGhee, and F. K. McKinney. 2006. Source code for theoretical morphologic simulation of helical colony form in the Bryozoa. *Palaeontologia Electronica* 9, no. 2, Article 7. http://palaeo-electronica.org/paleo/2006_2/helical/index.html.

3 LIFE'S EVOLUTIONARY HISTORY

Is it Determinate or Indeterminate?

Karl J. Niklas

Evolution may be determined—that is, completely caused in a materialistic way—and yet not rigidly predetermined from the first as to the course it was to follow. An equation can have multiple solutions, and yet each solution is determined by the equation.

George Gaylord Simpson

Introduction

Most biologists agree on the major trends in evolutionary history (e.g., Smith and Szathmáry 1995; Futuyma 1998; Strickberger 2000). Prebiotic replicating molecules were replaced by compartmentalized populations of independent genes that were replaced by chromosomes in unicellular prokaryotic cells that ultimately evolved into eukaryotes with the capacity for sexual reproduction and multicellularity, followed by the emergence of cell- and tissue-specificity (see Conway Morris 1998) (Fig. 1). Across each of these major evolutionary transitions, average body size increased, as did ecological specialization, particularly after multicellular organisms made their first appearance.

Yet, at finer levels of resolution, each of these transformations appears to be the statistical summation of numerous smaller trends,

Figure 1. Major evolutionary transitions in order of occurrence (top to bottom of list). List is not inclusive. Adapted from a variety of sources.

Replicating Molecules	⟶	Populations of Molecules in Compartments
Independent Genes	⟶	Chromosomes
Unicellular Prokaryotes	⟶	Asexual Clones ; Cell Walls
Multicellular Prokaryotes	⟶	Cellular Specialization; plasmodesmata
Unicellular Eukaryotes	⟶	Sexual Populations
Multicellular Eukaryotes	⟶	Tissue Specialization
Aquatic Multicellularity	⟶	Terrestrial Multicellularity

some of which have very different directions (Raup 1978, Jablonski 2000). Depending on the lineage (or the time interval) examined, body size may oscillate randomly or monotonically increase or decrease. Likewise, the degree of ecological specialization may vary over the long history of an individual lineage or large clade. Many specific examples can be drawn from the fossil record to support each of these contentions. But each reveals that the recognition and diagnosis of what may be called an evolutionary "trend" depends on our particular taxonomic (and temporal) perspective, as well as on the yardstick with which evolutionary change is measured (e.g., catalytic specificity, cellular organization, body size, anatomical or morphological details, and ecological specialization).

In this sense, the broad patterns evident in evolutionary history are fractal-like—they depend on the scale of measurement. Much like the length L^ε of a coastline depends on the length ε of the yardstick used, the meaning of a measurement of evolutionary direction only makes sense in the context of the size and nature of the yardstick used to measure it. In very general terms, however, L^ε increases as ε decreases so that very small scales of measurement tend to reveal considerable order. Indeed, as ε goes to zero, the relationship between the length of a coastline and the length of a yardstick is given by the equation $L_\varepsilon \sim \varepsilon^{-\alpha}$, where α (the fractal dimension) indicates that all the measurements are proportionally self-similar sets (Mandelbrot 1983).

Analogies can be misleading. But the foregoing comparison draws attention to the importance of quantifying and understanding self-similar sets in biology. That these sets exist is not in doubt. Numerous studies reveal "invariant" fractal-like scaling relationships between body mass and a vast array of physiological, phenotypic, and ecological features, ranging from resting metabolic rates and intracellular chemical concentrations to the architecture of circulatory systems and frequency distributions of species in communities (Brown and West 2000). These trends, some of which span seventeen orders of body size across prokaryotes and unicellular and multi-cellular aquatic and terrestrial eukaryotes (Niklas 1994a; Niklas and Enquist 2001), attest to the fact that some very fundamental phenomena underlie the entire fabric of life as we know it. That these phenomena are just as important to our understanding of evolutionary history as they are to com-prehending present-day ecology is an unequivocal fact of life. That they reflect a level of convergence that is unparalleled elsewhere in biology is also self-evident.

Contrasting Worldviews

Although interesting in their own right, it is unfortunate that the man-ifold fractal-like scaling relationships evident in modern-day organisms have no intrinsic directionality. Each is a summation of numerous "states-of-being"—snapshot views of extant organisms that have survived the gauntlet of natural selection and episodic mass extinction. To understand the evolutionary implications of self-similar biological sets, they must first be strung onto an evolutionary time line. Only then can we hope to inter-pret their historical implications.

Ordering the major groups of extant organisms along time's arrow is not a particularly contentious issue. It is generally acknowledged that bac-teria predate the eukaryotes (Smith and Szathmáry 1995; Knoll 2003), that unicellular life forms (protists) evolved before multicellular ones, that aquatic organisms predate terrestrial ones, and that nonvascular land plants predated the vascular plants. In this sense, there is a clear direction-ality to life's history.

Yet interpreting the emergent patterns of major evolutionary tran-sitions has proven contentious, particularly when different scholars use different yardsticks with which to measure them. Consider the polarized

Weltanschauung of Christian De Duve and Stephen Jay Gould. Both of these biologists espouse the existence of evolutionary patterns but for very different reasons. De Duve (1995) argues in favor of a clear directionality, at least in terms of the evolution of structural, catalytic, and informational molecules—a trend that goes from functionally general and inefficient biochemical reactions to progressively more specific and efficient ones. According to this view, evolutionary patterns emerge from orderly adaptive molecular modifications and innovations that ultimately translate into the familiar macroscopic world of the phenotype. According to this view, life's humble abiotic beginnings shaped the metabolic and genomic landscapes of aquatic and terrestrial organisms from the Precambrian to the Recent.

Gould (1989, 1996) also sees patterns in life's history but argues on morphological grounds that they are largely the result of unpredictable contingent events. Historical accidents ranging from developmental quirks early in the ontogeny of ancestors to global catastrophes throughout the history of life are argued to have far more influence than the adaptive role of natural selection. To be fair, Gould does not reject natural selection as a player in the evolutionary theater. But he does attempt to significantly diminish its role. In his worldview, natural selection merely confines the spread and accumulation of variance as species "diffuse" randomly from "life's left wall" (defined by the biology attributes of the first prokaryotes).

I would argue that these polarized views resonate with the analogy of measuring a coastline with different yardsticks. Indeed, I will argue further that De Duve and Gould are measuring two very different coastlines. De Duve's measuring stick is the molecule, and his coastline is the historical sequence of molecular evolution inferred largely from the physical rules and processes governing chemical reactions. Gould's measuring stick is the phenotype, and his coastline is the inferred historical sequence of morphological transformations. These fundamental differences in perspective and the conclusions that result from them are logical. De Duve sees fractal-like patterns and self-similar sets as the result of ordered and predictable molecular verities. In contrast, Gould sees continua and diffusion as the result of random historical events.

Certainly, determinism at the level of molecules does not preclude indeterminism at the level of phenotypes. The storage and retrieval of molecular information and its translation into enzyme-mediated catalysis and the emergent feature called "metabolism" are clearly governed by unalterable and, thus, unavoidable physical laws and processes. Arguably, the influence of these laws holds across the various higher levels of biological organization to the level of the phenotype, as attested by the abundance of convergent life forms in phyletically very different lineages. If homeomorphy is commonplace in life's history, determinism at the level of how environments dictate the external form and internal structure of animals and plants (by virtue of the operation of physical principles) must be as well.

Yet it is also clear that random or highly unpredictable events operating at the level of the genome to that of the ecosystem have played a significant role in evolutionary history. Consider the number of theoretically possible genotypic variants that a single mating pair of animals or plants can produce. If a single parent has N number of genes and each gene has two alleles, that parent can produce 2^N genetically different gametes (sperm or eggs) because genetic recombination is random. Because each mating event involves two parents, each mating pair can produce 4^N different genetic combinations, Assuming that a species has 150 genes, a single set of parents can produce 10^{90} different genotypes. This number exceeds the estimated number of atoms in the known universe, that is, 10^{80} (Hawking 1988).

Importantly, the number of genomic variants that any mating pair can produce is astronomically smaller (because organisms have a finite life expectancy and resources are always limiting). Thus, the genetic variants that a mating pair (or an entire population) of sexually reproductive organisms produces are a random sample (and infinitesimally small fraction) of what is possible. By the same token, many of those genomic possibilities that are realized (those that survive to birth) are expunged by natural selection. This extirpation can increase a population's fitness, but it can also operate randomly as the result of unpredictable abiotic events with differing mortality effects and temporal periodicity (e.g., local flooding, regional volcanic upheaval, and hemispheric asteroid impacts).

Life's Moving Left Wall

Certainly, at one level, it is easy to reconcile the polarized views of De Duve and Gould—traditional evolutionary theory canonically supports the duel roles of selection and random events in shaping the history of life. But can a closer agreement be found? Here, I argue that there is. If evolutionary history is fractal-like, there is an additional way to bring these two perspectives into closer accord.

My starting point is the recognition that large clades have rarely abandoned those attributes that define them. Clades may go to extinction (and even pseudoextinction). But they rarely if ever de-evolve. Prokaryotes have not been observed to devolve into protocells or independent replicating molecules. No eukaryotic lineage is known to have given rise to a prokaryotic one. Multicellular organisms rarely become unicellular. Evolution may be indeterminate and diffusive after each major transition, but the available evidence indicates that life's left wall is not stationary. With each new evolutionary innovation, it has been redefined—and it has moved to the right (see Niklas 1997; Knoll and Bambach 2000).

A few examples suffice to illustrate this point.

Consider first the allometry of prokaryotic and eukaryotic body size, particularly the relationship between surface area and volume, which influences the ability of organisms to exchange mass and energy with their environments. When data from representative prokaryotes and unicellular and small (free-living) multicellular eukaryotes are examined (Niklas 1994a, 1997), we see obvious differences in cell size frequency distributions and mean cell or body size (Fig. 2). Specifically, the mean (standard error) cell volume of prokaryotes is 363 + or − μm^3, whereas that of unicellular and small, free-living multicellular eukaryotes is on the order of 47,000 14,000 μm^3 and 5,000,000 + or − 2,400,000 μm^3, respectively. These differences in body size are hardly surprising. But consider the variance in cell shape within each grade of organization (as revealed by plotting body volume vs. length) and the extent to which species deviate, on average, from life's protocell left wall (a sphere, shown by a diagonal line in Fig. 3 A). Within each grade of organization, the measurements of some species fall on this line. But as we pass from one grade to the next, more and more species fall away from the left wall because of differences in cell or body shape. This trend is

Figure 2. Size-frequency distributions for representative prokaryotes (A), unicellular eukaryotes (B), and small, free-living eukaryotes (protists) (C). Species include photoautotrophs and heterotrophs in each category. Data taken from Niklas (1997).

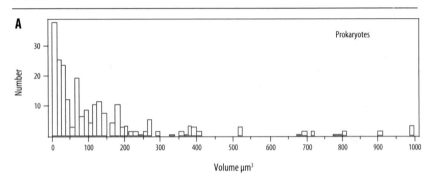

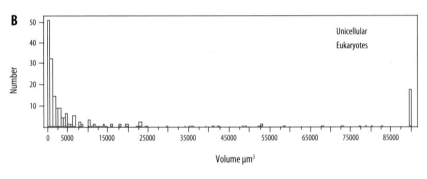

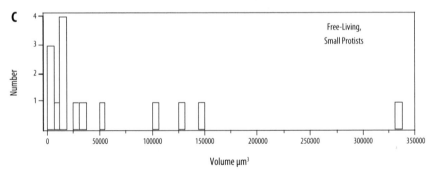

easily quantified by the quotient of body length and diameter Q for each of the three grades of cellular organization. For prokaryotes and unicellular eukaryotes, Q respectively equals 1.82 + or − 0.61 and 3.63 + or − 1.53; for multicellular eukaryotes, Q equals 12.7 + or − 3.20.

That this trend is not mere "passive diffusion" is revealed by the effect of body elongation on surface area and volume relationships. Provided that geometry and shape are conserved across entities differing in size, surface area always remains proportional to the ⅔-power of volume (Niklas 1994a). Larger spheres, thus, invariably have proportionally smaller surface areas than their smaller counterparts. However, if shape can be altered as size increases, the ⅔-power "rule" can be broken. And if shape and geometry are altered simultaneously and independently, the ⅔ "rule" becomes irrelevant.

That shape and geometry have been altered across the prokaryotic to unicellular eukaryotic transition (and again across the unicellular to multicellular eukaryotic transition) is evident when surface area is plotted against volume for the three grades of cellular organization (Fig. 3 B).

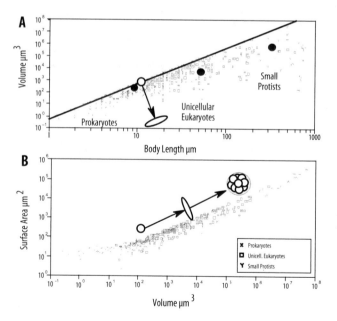

Figure 3. Log-log plots of body size (volume) versus body length (A) and body surface area versus body size (B) for data plotted in Figure 2 (see insert for key for grades of cellular organization). Volumes calculated assuming a spheroidal (body) geometry. Diagonal line in A denotes spherical body geometry; data points falling away from this line are progressively elongate (indicated by arrow in inserted diagram).

Inspection of the resulting relationship indicates a log-log nonlinear relationship across prokaryotes to unicellular eukaryotes. This nonlinearity indicates that cell shape has changed. Additionally, the slope of the log-log plot across unicellular and multicellular eukaryotes is not ⅔, but rather very near ¾, which indicates that, once again, body shape (and geometry) has changed across the unicellular to multicellular eukaryotic transition.

The adaptive benefits conferred by these changes are clear. Surface area influences the ability of an organism to intercept radiant energy, absorb nutrients, eliminate wastes, etc., whereas body volume provides a gauge, albeit crudely, of the metabolic demands for nutrients and the production of wastes. By amplifying surface area with respect to volume, growth can be maximized. The trends shown in Fig. 3 indicate that this happened during the pro- to eukaryotic transition and again during the uni- to multicellular transition. Life's left wall has moved to the right not because of random walks or passive diffusion but as a consequence of adaptive evolution.

The colonization of land by plants provides another example. The most ancient land plants had diminutive, nonvascular stem-like axes composed almost entirely of parenchyma (Taylor and Taylor 1993; Stewart and Rothwell 1993). During the Silurian and early Devonian, plants evolved taller stems with primary xylem and phloem. Plants with woody tree-sized stems evolved by the end of the Devonian. The descendents of many of these organisms survive today, affording an opportunity to examine the consequences of stem-tissue innovations (from parenchyma to primary xylem to wood) on plant stature. Using the mechanical properties of each of these stem tissues, the maximum height to which any stem can grow can be calculated. Assuming that stems are composed of only one tissue type, engineering theory shows that maximum height must scale as the ⅔-power of stem diameter (McMahon 1973; Niklas 1992). Engineering theory also shows that, with increasing tissue stiffness, the Y-intercept of this line will be ratcheted up. Thus, the maximum heights for stems with equivalent diameters but composed of different tissue types are depicted as a family of parallel lines with a slope of ⅔ (Fig. 4).

Plotting actual data for plant height and diameter against these parallel lines shows that the left wall has moved upward after the evolutionary appearance of each new plant group relying on a stiffer tissue type

Figure 4. Log-log plot of plant body length (height) versus basal stem diameter for representative data from moss, pteridophyte, and dicot tree sporophytes. Diagonal lines denote the maximum height to which any stem with a specified diameter can grow vertically before it becomes elastically unstable assuming that stems are composed of parenchyma (par.), primary xylem and phloem (xylem), or wood. Note the data for trees parallel that of the line predicted for wood because these data are for the largest specimens reported for each species and because the trunks of trees are made mostly of wood. Data taken from Niklas (1994a).

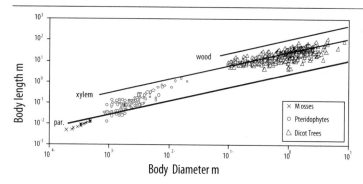

for mechanical support. By the same token, within the mechanical constraints imposed by each tissue type, the species within individual lineages have diffused toward the upper limits set by their ancestral stem tissue composition (Fig. 4). Certainly, for some lineages, ecological specialization has resulted in a reduction in stem stiffness (and stature). But the general anatomical trend in land plant evolution is clear and, when viewed with mechanical and hydraulic requirements in mind, very predictable.

The Frequency of Homeomorphy

There is one more example of life's moving walls that bears on the debate over the relative importance of random versus nonrandom factors and their influence on evolutionary history: the allometry of body mass and length (Fig. 5).

Across twenty-two orders of magnitude of mass and eight orders of magnitude of length, a log-log linear relationship exists when body mass is plotted against body length. This linear relationship holds true for unicellular and multicellular eukaryotic life forms, and it is indifferent to both

Figure 5. Log-log plot of body mass versus body length for unicellular and multicellular species. A. Data distinguished between plants and animals. B. Data distinguished between aquatic and terrestrial organisms. Data taken from Niklas (1994b).

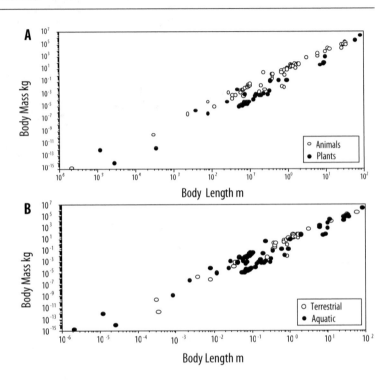

habitat and the otherwise fundamental metabolic distinctions between photoautotrophs and heterotrophs because the slope of the log-log regression curve for mass versus length neither differs statistically between aquatic and terrestrial organisms nor between plants and animals (Niklas 1994b). Across all of the diverse species examined thus far, the slope is three. Thus, body mass M scales as the third power of body length L, i.e., $M \propto L^3$.

This scaling relationship is called geometric self-similarity. It invariably emerges if body diameter D scales one-to-one with length (i.e., $D \propto L$) and if bulk tissue density ρ is more or less invariant across life's manifold phenotypic expressions (because $M = \rho D^2 L$).

Mathematically, geometric self-similarity is a trivial condition. It is achieved by any series of more or less cylindrical objects differing in size when diameter increases in direct proportion to length. On this basis, one can argue that adaptive evolution is not required to achieve geometric self-similarity because it is "so simple" (in comparison to alternatives like elastic or stress self-similarity). Biologically, however, the proportional relationship $M \propto L^3$ is highly adaptive because it facilitates the exchange of essential nutrients with the fluid in which an organism lives, enhances the ability to compete for nutrients and space, and effectively promotes movement on land or through viscous liquids or soils (Wainwright 1988). Additionally, within the size ranges occupied by land plants, it promotes the dissemination of propagules (seeds and fruits), whereas, in water, it expedites reaching the upper sunlit portions of the water column while remaining attached at the other end to a substrate. Certainly, some organisms deviate significantly from geometric self-similarity. Sea urchins, box turtles, barrel cacti, and *Lithops* are excellent examples. But it is reasonable to argue that the great number of plant and animal species that scale geometrically in tandem with the manifold other mass-to-length scaling relationships that could but did not emerge provide evidence that natural selection rather than random events is responsible for the trend shown in Fig. 5.

Accordingly, I argue that the $M \propto L^3$ scaling relationship is one of the best examples of convergent evolution and that it provides strong evidence that homeomorphy is commonplace. In this context, evolutionary history was shaped (literally) far more by the operation of natural selection than by random events. Clearly, as a consequence of evolutionary changes in body size (which abide by other but interrelated scaling relationships), the representatives of various plant and animal lineages have "migrated" diagonally through the corridor of the $M \propto L^3$ relationship. In this limited sense, there has been "diffusion" in the trajectory of animal and plant body-shape evolution. Importantly, however, the boundary conditions established by natural selection have confined plants and animals laterally within a narrow corridor, which has been rarely breached (and only then as a consequence of adaptive evolution).

Concluding Remarks

As a plant biologist, I cannot resist pointing out that well over 90 percent of all visible life is capable of fabricating its living substance from carbon dioxide, oxygen, water, and minerals by virtue of harnessing the energy of sunlight. If biologists feel compelled to argue about the "relative frequency" of a particular evolutionary phenomenon, they would do well to consider this fact and examine plants to resolve the debate. This admonition seems particularly relevant to arguments about the relative frequency of convergent evolution because this phenomenon is ubiquitous among extant plant species and evident at every turn in the history of plant life.

The reason for the prevalence of homeomorphy among plant life forms is simple. Because of their basic metabolic requirements, plants are photovoltaically driven chemical factories. By the same token, with the exception of unicellular and colonial species with flagellated cells, plants are immobile and sedentary organisms that lack any neurological counterpart. As such, their physiology and the manner in which it is structurally housed and spatially deployed are tightly constrained by the laws of chemistry and physics. Across all of plant life, the phenotype is a structural solution to reconciling the physical requirements for gas exchange, light interception, nutrient adsorption, and coping with externally applied mechanical forces. Engineering theory reveals the existence of only a few phenotypes capable of reconciling all of these requirements simultaneously, whereas the fossil record shows that these "optimizing" phenotypes have evolved repeatedly and independently in the majority of lineages. Thus, the things we call *leaves* have evolved independently in the mosses, liverworts, lycopods, and ferns, just as roots have evolved in at least three very different lineages (lycopods, horsetails, and ferns). By the same token, much of plant reproduction is dictated by physical laws and processes, as attested to the large number of wind-pollinated species whose reproductive structures comply in every respect with what aerodynamics dictates.

For these reasons, it seems appropriate to end this paper with a quote from D'Arcy Thompson:

So the living and the dead, things animate and inanimate, we dwellers in the world and this world wherein we dwell—π α ντα γα μαν τα γιγνωσκ ο μενα—are bound alike by physical and mathematical law.

References

Brown, J. H., and G. B. West, ed. 2000. *Scaling in biology.* Oxford: Oxford University Press.

Conway Morris, S. 1998. *The crucible of creation: The Burgess Shale and the rise of animals.* Oxford: Oxford University Press.

De Duve, C. 1995. *Vital dust: Life as a cosmic imperative.* New York: Basic Books.

Futuyma, D. J. 1998. *Evolutionary biology.* 3rd ed. Sunderland, MA: Sinauer Assoc., Inc.

Gould, S. J. 1989. *Wonderful life: The Burgess Shale and the nature of life.* New York: Norton.

_____. 1996. *Full house: The spread of excellence from Plato to Darwin.* New York: Harmony Books.

Hawking, S. W. 1988. *A brief history of time.* New York: Bantam.

Jablonski, D. 2000. Micro- and macroevolution: Scale and hierarchy in evolutionary biology and paleobiology. In *Deep time: Paleobiology's perspective*, ed. D. H. Erwin and S. L. Wing, 15–52. (Supplement to volume 24 of the journal *Paleobiology*). Lawrence, KS: The Paleontological Society.

Knoll, A. H. 2003. *Life on a young planet.* Princeton, NJ: Princeton University Press.

Knoll, A. H., and R. K. Bambach. 2000. Directionality in the history of life: Diffusion from the left wall or repeated scaling to the right? In *Deep time: Paleobiology's perspective*, ed. D. H. Erwin and S. L. Wing, 1–14. (Supplement to volume 24 of the journal *Paleobiology*). Lawrence, KS: The Paleontological Society.

Mandelbrot, B. B. 1983. *The fractal geometry of nature.* New York: W. H. Freeman.

McMahon, T. A. 1973. The mechanical design of trees. *Science* 233: 92–102.

Niklas, K. J. 1992. *Plant biomechanics.* Chicago: University of Chicago Press.

_____. 1994a. *Plant allometry.* Chicago: University of Chicago Press.

_____. 1994b. The scaling of plant and animal body mass, length, and diameter. *Evolution* 48: 44–54.

_____. 1997. *The evolutionary biology of plants.* Chicago: University of Chicago Press.

Niklas, K. J., and B. J. Enquist. 2001. Invariant scaling relationships for interspecific plant biomass production rates and body size. *Proc. Natl. Acad. Sci. (USA)* 98: 2922–27.

Raup, D. M. 1978. Cohort analysis of generic survivorship. *Paleobiology* 4: 1–15.

Smith, J. M, and E. Szathmáry. 1995. *The major transitions in evolution.* Oxford: W. H. Freeman.

Stewart, W. N., and G. W. Rothwell. 1993. *Paleobiology and the evolution of plants.* Cambridge: Cambridge University Press.

Strickberger, M. W. 2000. *Evolution.* 3rd ed. Sudbury, MA: Jones and Bartlett Publishers.

Taylor, T. N., and E. L. Taylor. 1993. *The biology and evolution of fossil plants.* Englewood Cliffs, NJ: Prentice Hall.

Wainwright, S. A. 1988. *Axis and circumference.* Cambridge, MA: Harvard University Press.

4 EVOLUTION AND CONVERGENCE

Some Wider Considerations

Simon Conway Morris

Introduction

Of all the sciences, perhaps the richest in metaphors is biology. In large part, these terms of expression reflect a belief in the indeterminacy, the open-endedness, the sheer unpredictability of the evolutionary process. Among the most familiar terminologies will be those of a "blind watchmaker" (Richard Dawkins), a "tinkerer" (Jacques Monod), or Stephen Jay Gould's conceit of "re-running the tape of life." Each addresses a somewhat different aspect of evolution, but all these are consistent with the notion that both the process and, more importantly, the end result are random and accidental. These, and similar, tags reflect also a variety of agendas, including those of atheism and relativism, but unconsciously they pose the paradox that, even if the processes involved are blind, somehow we not only find ourselves as creatures in possession of meaning but, as often as not, are entirely awestruck at the complexities of resultant systems, be it in the utter intricacies of a bacterial cell or a singing human. Indeed, as our knowledge, especially of biochemistry and protein function, continues to expand, so at least my sense of amazement can only grow. If the watchmaker is blind, he has an unerring way of finding his way around the immense labyrinths of biological space. And even if he doesn't know where he is going, does He still know?

At first sight, this antithesis between the naturalistic world picture and a metaphysic that ultimately leads to a teleology would seem to be a conjunction of false opposites, a sort of category mistake. In many respects the study of evolution seems unequivocally to support the world picture of Dawkins, Gould, and Monod—or so it would appear. Consider, for example, three key concepts, all central to the evolutionary synthesis: mutations, mass extinctions, and adaptive radiations. Any textbook on evolution failing to address three concepts would be badly amiss, but, in each case, important qualifications need to be made that show there are subtleties of interpretation that are less often explored, at least in some elementary textbooks on evolution.

Mutations are widely construed as the motor of evolution, effectively random and whose end results are mostly for the worse. If chance—a cosmic ray?—and disadvantage are the norm, then it is easy to see why evolution is so often seen as a largely fortuitous process. Yet there is considerable evidence that the cards are actually very much stacked in the opposite direction, and that evolvability itself is a selected trait (e.g., Caporale 2003a,b; Earl and Deem 2004; see also Perfeito et al. 2007). The identification of hypermutation and site sensitivity, especially in response of pronounced environmental changes, strongly suggests that, when the cards are dealt across the table of life, aces and kings appear with alarming frequency. The concept of evolvability is an important shift away from the prior emphasis on randomness. To be sure, evolution is based on genetic change, but there are pervasive biases. Caporale (2003a, 467) has gone so far as to write, "Indeed, some potentially useful mutations are so probable that they can be viewed as being encoded implicitly in the genome," and she notes further how "challenges and opportunities tend to recur, [so] a response that is better than random can be favored by selection" (468).

What of mass extinctions? Here, too, the role of chance and accident, combined with lurid descriptions of the catastrophic circumstances descending onto an unsuspecting world, has provided a powerful impetus to evolutionary thinking in terms of the radical, and unpredictable, redirection of the history of life. In promulgating these ideas, the late S. J. Gould may have been one of the most strident, but the general notion of accident and redirection has received widespread, if uncritical, approval. The focus of attention, unsurprisingly has been the end Cretaceous (K/T) event. This is because of both the evidence for a catastrophic bolide impact and the crippling of reptilian diversity. The latter, of course, gave the ecological baton to the birds and mammals, which, in post K/T time, rapidly radiated into numerous niches. In a counterfactual world, where the asteroid missed, the birds and mammals would have literally remained in the shadow of the great reptiles. In simple terms, the argument runs as follows: No impact? Then, no mammalian radiations. No radiation? Then, no primates. No primates? Then, no apes, let alone us.

Faced with this vertiginous prospect of potential nonexistence were it not for a falling star, two items tend to be overlooked. First, of the other four big mass extinctions in the Phanerozoic (i.e., from the beginning of the Cambrian, and so an effective fossil record), the net result of three of them (end-Ordovician, late Devonian, and Triassic) was muted. What succeeded did not differ so greatly from what preceded. Second, and more significantly, a good argument can be made that even those two mass extinctions, that is, K/T and end-Permian, that are hugely catastrophic, nevertheless, only serve to accelerate (or postpone) the course of evolution, but seldom, if ever, do they irrevocably divert the overall path. Thus, in the case of the K/T event, let us imagine the counterfactual world where the bolide sails harmlessly by as a bright light in the sky of a June evening (why June? See Wolfe 1991). This parallel world would, we may assume, show the same climatic history as the Earth, including, of course, the onset of major glaciations from about thirty-five million years ago. How would evolution respond to this massive environmental challenge? There is little doubt that the warm-blooded birds and mammals that, recall, were coexistent with the dinosaurs would have seized the opportunity, rapidly diversifying in the temperate and polar zones. The tropics, to be sure, would remain the abode of the great reptiles, but nearer the poles we can predict that the diversification of the warm-blooded groups would see the emergence of complex organizations, including vocalization (e.g., Beckers et al. 2004), tool making (e.g., Hunt and Gray 2003), social play (e.g., Diamond and Bond 2003), and cooperative hunting (e.g., Bowman 2003). Why? Because all these have evolved in the birds, and so quite independently of the mammals, as indeed has warm-bloodedness itself (see Farmer 2003). So, given these examples of convergence, it seems as likely in this counterfactual world that sooner rather than later hunter-gatherers would have emerged. The mass extinction of the dinosaurs would then have been under way, perhaps thirty million years behind schedule in comparison with the real world.

What, then, of adaptive radiations? If mutations and mass extinctions, in their very different ways, have effects considerably less accidental than popularly imagined, are not adaptive radiations the exemplar of evolution following all or every available path? At first sight, indeed so. Take the hominids: here we are, magnificent bipeds and direct lineal descendants of shambling chimp-like figures that, about six million years ago, hovered nervously at the forest edge. A simple lineage? Not really, because, despite the disagreements as to the numbers of hominid species, overall the bush of diversification shows numerous ramifications and, unsurprisingly, such dead ends as the robust paranthropoids, somewhat later the Neanderthals, and even astonishingly "microcephalic" miniaturized island-dwellers (Brown et al. 2004; Lahr and Foley 2004; Morwood et al. 2004; Tocheri et al. 2007). The major qualification, however, is that, despite this, the history of hominids is riddled with parallelisms, be they megadonty, symbolic culture, quite possibly tool use, and maybe even key features such as bipedality. Nor are hominids in this context any sort of exception. The rule here, as everywhere else, is that any phylogenetic diversification is laden with so-called homoplasies, that is, the recurrent and independent emergence of given features.

This introductory critique suggests that evolution may be considerably less random than is often supposed. My principal argument to extend this conjecture is the prevalence of evolutionary convergence. I will return to this topic in more detail below, but, at this juncture, it is also worth drawing attention to an interesting series of biological rules. To be sure, their universality remains a source of debate, but, at the least, there is an argument that in the project "What Life Is That? A Space-Farer's Guide to Extraterrestrial Biospheres," the introductory chapter will draw attention to such rules as those of metabolic (Savage et al. 2004; but see Bokma 2004) and allometric (West et al. 1999) scaling, as well as those that pertain to flight (Maurer et al. 2004; Nudds et al. 2004) and swimming (Linden and Turner 2004), not to mention global ecological constraints such as latitudinal gradients (Hillebrand 2004). Turning to the chapter on plants, no doubt we will be forcibly reminded of such universals as rain-use efficiencies (Huxman et al. 2004), altitudinal tree lines (Jobbagy and Jackson

2000), biomass production (Niklas and Enquist 2001), ecological context of productivity versus disturbance (Grime 2001), heterospory (Bateman and DiMichele 1994), propagule microarchitecture (Hemsley et al. 2004), the golden angle in phyllotaxis (King et al. 2004), flexural stiffness (Niklas 1991), and leaf structure (Reich et al. 1998; Boyce and Knoll 2002). All these are convergences found on Earth, but so intimately are they tied to physico-chemical constraints that it is difficult to believe they will not be found everywhere.

It seems, therefore, that the study of evolution, and especially of evolutionary convergence, will provide us with the outline of a "map." In some ways, it will be a strange sort of cartography because the map shows mostly the evolutionary destinations. How you get to a particular destination, say, humanoid intelligence, is distinctly less important than the end result. If, as I argue, evolution has some fundamental predictabilities, then is not the vocabulary of surprise associated with the documentation of evolutionary convergence in itself all the more surprising? Almost invariably the words tend towards adjectives of stupefaction: *astonishing*, *astounding*, *remarkable*, *striking*, even *uncanny* and *stunning*, are all stock-in-trade responses. As I have pointed out elsewhere, although pronounced by loyal Darwinians, these exclamations seem to reveal a sense of unease. This, I conjecture, is at the least reflecting a hesitancy as to evolution's having a degree of directionality and, perhaps in the more alert investigator, their worst fears of the reemergence of a telos.

These remarks may serve to persuade the distinterested reader that there is a fundamental ground to biology, that not all is possible, and indeed perhaps nearly everything is impossible. Nevertheless, this would not necessarily allow us to predict specifics of outcome, such as singing or humanoid intelligence. Yet the evidence is that these are blatantly convergent, as is explored in my book *Life's Solution* (Conway Morris 2003). Rather than reiterate the data and conclusions set out there, it should be more helpful to look as to how this field is developing and, more importantly, what some of the wider implications might be.

In the immediate context of identifying specificities that might refine the construction of a "map of life," there are four areas I will address briefly: experimentation, molecular convergence, biological properties such as "mammalness," and brains and intelligence.

Experiments in Evolutionary Convergence

Studying evolution in the laboratory has, of course, a long history, but the specific context of identifying precise trajectories and corresponding convergence has received much less emphasis. Nevertheless, a number of interesting results is available. Given the advantages of rapid generation times, genomic data, ease of laboratory manipulation, and specification of environmental parameters, it is not surprising that the emphasis has been on viral (e.g., Bull et al. 1997; Wichman et al. 1999) and bacterial (e.g., Nakatsu et al. 1998; Cooper et al. 2003) systems. Despite the diversity of approaches and protocols, it is striking how often the evolutionary trajectories are far from random. In particular, there can be pronounced recurrences of outcome in response to adaptive circumstances, although, significantly, as often as not the path to a similar, if not identical, solution is interestingly different. The extent to which the experimental work can be applied to higher organisms remains to be seen, but there are at least two avenues worth considering. The first is a simple extension of the existing work, as in the case of the experimental evolution in fly (e.g., Matos et al. 2004). The second is considerably more ambitious and not without its attendant risks. This possibility is explored by Bennett (2003), who inquires as to the extent of "preferred pathways" and the likelihood of evolutionary recurrence and thereby the inevitability of the process from a given starting point and adaptive challenge. The potential risk lies in the fact that, to explore this possibility, Bennett suggests the adoption of genetic engineering methods in order to construct organisms with necessary prerequisites.

A rather different approach would be a systematic survey of phylogenies to see what, if any, patterns of convergence occur. In this context, an important contribution is made by Wagner (2000), who documents how homoplasy, the term used by cladists to denote the same character evolving independently, becomes increasingly frequent as a clade repeatedly diversifies. Wagner (2000) shows that the available "space" for diversification becomes "exhausted." Faced with an eternal return, convergence, therefore, becomes ever more likely. This approach has two important qualifications. First, the very identification of homoplasies may have an element of circularity, given that any phylogeny is based on the fact that, to be operational, at least two taxa must possess the same character. But is it the same because the character in question was possessed by the common ancestor, or is the "same" because it evolved independently? Even the question of "independence" is begged with the growing realization that characters are effectively latent, if not cryptic, with their appearance dictated by particular genes in particular adaptive circumstances (e.g., Cresko et al. 2004; Mundy et al. 2004; Sucena et al. 2003). The prospect of characters "flickering" in and out of phylogeny is only one of a set of complications that bedevil the innocent cladist (if such an unlikely figure exists). In terms of other difficulties, particularly important is so-called concerted convergence, whereby a series of characters are so interdependent that they evolve as an integrated unit.

In terms of exploring phylogenies in ways that might further refine our understanding of convergence, it will be useful to look to general biological properties that have clearly evolved independently but in markedly different adaptive contexts. Here, too, general principles might emerge. The range is wide and could encompass: coloration, miniaturization, monogamy, music, parasitism, respiration, sensory perceptions, simplification, social play, sound production, swimming, thermoregulation, tool making, viviparity, etc. Such an approach has several interesting aspects. For example, it is neutral as to the directions in which evolution "chooses" to run: reversals are just as interesting as the emergence of further complexities (e.g., Porter and Crandall 2003). In addition, the range of investigations is substantial and ranges from the structure of communities and societies to the molecular level. It is to this latter topic that we can now turn briefly.

Molecular Convergence?

At first sight, the prospect of molecular evolution leading to convergence would seem to be highly improbable. This is simply because of the combinatorial immensities of sequence "space"; the immense number of alternatives far exceeds the number of atoms in the visible universe, let alone the number of seconds from the big bang. To find, therefore, two proteins that on all other evidence were not closely related but had an identical sequence of amino acids would be deeply suspicious. Even so, constraints might still be stronger than generally realized, be they in terms of structure (Denton et al. 2002) or functionality of fold sites (Axe 2004). It is also the case that, while examples of molecular convergence are rare, the number of examples is now growing rapidly, and it is likely that its frequency has been underestimated. Identification of molecular convergences has three important implications, two theoretical and the third more practical. Recognition of a molecular convergence potentially throws light on the possible evolutionary pathways and the nature of the functional intermediates. The importance of this in the context of "intelligent design" and claimed irreducible complexity will be self-evident. Intrinsic to the identification of convergence is the reasonable assumption that it reflects adaptation, and this applies with equal force to the many molecular examples now documented. It is, therefore, consistent with the growing list of examples for positive Darwinian selection and suggests that assumptions of neutral molecular evolution require stringent reassessment. The third implication, that of practicality, is that the ability to "navigate" to the same solution repeatedly has immediate relevance to the appearance of both insecticide resistance (e.g., Weill et al. 2004) and virulent diseases (e.g., Reid et al. 2000). Even existing setbacks in disease and pest control, with the rapid emergence of resistant strains, suggest that attempts to design new molecular configurations will require considerable subtlety.

Documentation of molecular convergence, let alone identification of general principles, is still at a primitive stage, and here it will only be possible to provide a few pointers. Available examples are highly disparate and, apart from the customary exclamations of surprise, rather seldom do the investigators seek to probe into the deeper constraints that might dictate molecular convergences. By this, I do not mean to disparage in any way the brilliant nature of the specific investigations. Rather, it is a question of looking toward more general principles, both the lure and pitfall of biology. For example, in documenting the extraordinary mimicry of bacterial proteins that enable them to "outwit" the defenses of the host cell, Stebbins and Galan (2001) reviewed the case of a bacterium (related to the plague agent) that employs a protein (aptly named "invasin") that out-competes the host's ability to use its own proteins (such as fibronectin) to bind the surface receptors ($\beta 1$ integrins). The overall protein structures of fibronectin and invasin are very different, but the key mimicry involves just three amino acid residues with almost identical positions in the binding site. To succeed in penetrating the host defenses, invasin must converge in this precise way; but, given this, why exactly is invasin able to out-compete fibronectin? The question is important not only because of disease control but because, however striking the similarities, the convergences are seldom precise. Thus, while evolutionary convergences will be central to delineating the "map" of life, it should never be forgotten that the evolutionary pathways to particular solutions have their own intrinsic interest. More importantly, in evolutionary convergence, there is sometimes evidence for a scale of effectiveness. To give a well-known example, the striking convergence between the camera-eyes of some cephalopods and the vertebrates is well known, but arguably those of the latter are "better" in such contexts as lens structure and extraocular musculature (see Conway Morris 2003). On the other hand, the position of the retina in the cephalopod eye is arguably superior to the inverted arrangement found in the vertebrates. Properly documented, these and other examples might give new insights into both evolutionary constraints and contingent factors.

As was noted briefly above, another aspect of molecular convergence is the repeated appearance of a character by the expression of a particular gene that is effectively "on demand" when a given adaptive need arises. Even more interesting is the evidence for the same character emerging but on the basis of different genes. Most of the present examples are relatively parochial, such as pigmentation patterns in flies (Wittkopp et al. 2003) or melanism in mice (Hoekstra and Nachman 2003). Of potentially greater significance is whether major components of body-plan construction, which in animals includes sensory organs, dorso-ventrality, segmentation, body cavities, and appendages, arose by repeated recruitment of particular developmental genes rather than by common ancestry, as is more usually supposed. Such a possibility is enhanced by the emerging evidence for convergent evolution in gene networks (e.g., Conant and Wagner 2003; Amoutzias et al. 2004; see also Weinreich et al. 2006). Here the prior expectation had been that networks were primitive and deployed increasingly widely by such processes as gene duplication. This certainly plays a role, but the evidence points as strongly to convergent emergence. For example, in one gene network (MIM-2) of 176 identified circuits in yeast, 168 had an independent ancestry (Conant and Wagner 2003). So, too, in a study of the basic helix-loop-helix (bHLH) protein family, Amoutzias et al. (2004) showed some striking parallelisms, such as between the so-called *Max* and *E2A-Arnt* networks, emphasizing "that there are similar restraints on the evolution of the different networks" (277) in this protein family.

Yet another way to explore the likelihood of alternatives is to decide why a "molecule of choice" is just that. Such is evidently the case for the opsins (Fernald 2000). Other molecules can serve to transduce photons, but, in the case of vision, opsins are the molecule of choice; in Fernald's words, opsins have "proven irresistible for use in eyes" (446). Similar arguments may apply to many other proteins. Even where there are alternatives, and here the elastic proteins spring to mind, the total is still very limited.

Respiratory proteins provide another instructive instance. The key example is haemoglobin (and the closely related myoglobin), which is found from bacteria to plants and animals. It is likely that these iron-globins are convergent (e.g., Wittenberg et al. 2002), yet their near ubiquity suggests a peculiar suitability for the transport and storage of oxygen. Note also that, although the roles are most familiar with respect to aerobic respiration, these proteins are as important in the exclusion of oxygen from sites of anaerobic metabolism. Are there alternatives? Certainly. Most notable in this respect is the copper protein haemocyanin, but this too is convergent with independent origins in the arthropods and molluscs (e.g., Immesberger and Burmester 2004). It is the third example, haemerythrin, that suggests haemoglobin (and to some extent haemocyanin) really are the molecules of choice. In animals, the distribution of haemerythrin is highly sporadic and simply inconsistent with any proposed phylogeny that would allow a common ancestry. This iron protein, however, also occurs in bacteria, and the most likely scenario is one of an ad hoc recruitment. Why a handful of annelids, as well as the brachiopods, priapulids, and sipunculans, have chosen haemerythrin is not known, but the point remains that, as a potential respiratory protein, haemerythrin is "available." Even though it too walks a metaphorical tightrope of functionality and, thus, represents one of "three solutions to a common problem" (Kurtz 1999), it seems that haemoglobin will nearly always win out.

The topics within molecular convergence will doubtless grow in the next few years. This is not only because of the intrinsic interest but also because of implications for such areas as the molecular "clock" (see Vowles and Amos 2004). Some indication of the scope of possibilities should come from a couple of following examples. As already indicated, given the functionality of proteins, and especially enzymes, the fact that the same solution has been arrived at independently should not be cause for surprise, even if it usually provokes just such exclamations. A striking example is carbonic anhydrase, a highly effective enzyme with the characteristics of a zinc atom in the active site and a two-step reversible reaction that hydrates carbon dioxide. Carbonic anhydrase is central to processes as diverse as photosynthesis, respiration, and biomineralization. It would be reasonable to assume this enzyme is very primitive and has been recruited as and when required. The reverse is the case. Carbonic anhydrase has evolved at least three times, and, although the enzymatic process is identical, the actual proteins are wildly disparate (see Tripp et al. 2001).

An even more surprising example comes from the eye. It has long been recognized that the crystallins, the proteins employed to confer transparency to the lens and cornea, have been recruited from multiple sources, most usually proteins involved with heat-shock or stress management. Much more remarkable is the documentation of the promoter sequence connected to crystalline expression in the scallop eye and such vertebrates as mouse and chick. The bivalve mollusc employs a Ω-crystallin, derived from an enzyme known as aldehyde dehydrogenase, while the vertebrates use an fA-crystallin, which is related to small heat-shock proteins. Unsurprisingly, given their quite separate origins, these two proteins have no sequence similarity, even though they achieve identical functions. What is much more noteworthy is that, despite these differences, the promoter sequence of either protein is convergent (Carosa et al. 2002). The moral is that, although there are many ways to see, one still ends up looking at the same thing.

Can We Define Biological Properties?

If evolution has a spectrum of discussion, and one end point is molecular, the other is the level of complexity that is a prerequisite of intelligences that will lead to self-knowledge and ethical action. In this context,

I will briefly consider three biological properties, specifically, mammalness, brains, and mentalities. All are imprecisely defined, and there are also areas of overlap. Nevertheless, investigation of such properties allows us to distance ourselves from the historical perspective, which, although integral to understanding evolution, may also serve to obscure deeper, possibly timeless, patterns.

The extent to which other groups approach mammals in terms of biological complexity has perhaps been underestimated, but such comparisons are sufficiently close as to invite the concept of "mammalness." Indeed, we can go further than this to suggest that this recurrence of the particular biological property of mammalness hints that this—and other—properties might be universals. So how might we begin to define *mammalness*? Somewhat surprisingly, the reptiles show a number of instructive examples. The best known instance concerns reproduction. A telling instance is the convergence in penis design between turtles and mammals (e.g., Kelly 2002, 2004). So too is the repeated evolution of viviparity, especially in lizards where the convergences extend to striking similarities of the placenta (e.g., Blackburn 1992; see also Flemming and Blackburn 2003). Somewhat less remarked upon is the degree of mammalness seen in the monitor lizards (Horn 1999). Thus, in terms of physiology, active locomotion (that can include bipedality), behavior (including courtship), hunting, and intelligence, the monitors are remarkably mammal-like. In addition, other lizards are noteworthy for their sophisticated social structures, notably the Australian skink *Egneria* (O'Connor and Shine 2003; Lanham and Bull 2004).

Given mammals evolved from reptiles, albeit not lizards, the emergence of mammal-like features in the lizards themselves is perhaps less surprising than might be thought, even if the convergences are remarkable in their range and sophistication. In other ways, however, the degree of mammalness shown in the birds is even more noteworthy. I have reviewed this area elsewhere (Conway Morris 2003) and here will touch only on the principal convergences. Among the most striking similarities are warm-bloodedness (Farmer 2003), which significantly shows a different physiological basis (see Schweitzer and Marshall 2001); parental care (Farmer 2000); sociality (Bond et al. 2003) that almost certainly extends to cooperative hunting (Bowman 2003); social play (Diamond and Bond 2003); sophisticated vocalizations that in parrots have speech-like properties

(Beckers et al. 2004) and in ravens includes cultural transmission (Eng-gist-Dueblin and Pfister 2002); music (Bottoni et al. 2003); and tool making, which ranges from the strategic placing of dung as beetle bait by owls (Levey et al. 2004) to crafted technologies made famous by the New Caledonian crows (Hunt and Gray 2003, 2004; see also Chappell and Kacelnik 2004). This impressive range of capabilities is underpinned by cognitive sophistication (Emery and Clayton 2004; Lefebvre et al. 2004), but it is important to stress that these abilities are based on a brain with a radically different ground plan to that of the mammals (see Rehkämper and Ziller 1991). There are two other contexts that are also worth mentioning when we consider the striking structural and functional similarities between birds and mammals. The first is that there is a range of capabilities, a spectrum of sophistication (e.g., Sol et al. 2002). As with birds, so with primates, and presumably cetaceans. The second point is that very complex behaviors may arise in specific adaptive contexts. In the case of the New Caledonian crows, for example, Kenward et al. (2004) suggest that the emergence of tool use is related to the search for protein-rich food on an island with no large mammals. One wonders what parallels might exist with early hominid evolution.

Convergent Mentalities?

The intelligence, if not the *chutzpah*, of the birds should surely serve to reignite our interest into not only the evolutionary origins of this strangest of biological properties, touching as it does on the mysterious question of consciousness, but also the intriguing question as to whether all intelligences could tend towards a similar end point. So far as primate-like intelligence is concerned, its independent appearance in birds and cetaceans—and note well on the basis of markedly different neural substrates—supports such a conjecture. What in my opinion are striking examples of convergence underpins that such features may well be evolutionarily inevitable but, as importantly, suggest that primate mentality (and its equivalent cetacean and corvid/psittaciform manifestations) is a real property, neither an epiphenomenon nor nominal. To help reinforce this point, consider two very different activities, both of which are also convergent. One is sleep, which is found in animals as far removed from vertebrates as insects (e.g., Kaiser 1988; Cirelli 2003), and perhaps more surprisingly crayfish (Ramón et al. 2004). In sleeping bees, spontaneous antennal movements (Sauer et al. 2003) beg the question of an equivalence to our rapid eye movement; do bees dream? Some animals evidently do not sleep, but there is little doubt that, where animals do sleep, it is essential for mental functions, including memory (e.g., Walker and Stickgold 2004; but see Vertes 2004).

If, in our unconscious, we reorder out worlds and possibly discover new ones, so another facet to our lives is self-identity. Given that even among wasps individuals can be recognized (Tibbetts 2002), this suggests yet another mental property that may be widespread and possibly convergent. This is not to deny that there are levels of self-identity, and, in this respect, a key, and very rare but still convergent, step is self-recognition in a mirror. This has evolved independently in the dolphins (Reiss and Marino 2001). Knowing this, it is less surprising that these animals also have considerable skills of gaze comprehension. Intriguingly, this ability probably arose not only because of their social structure but also sensory perception based on the use of echolocation rather than eyes (Pack and Herman 2004).

The evolution of intelligence leads to a series of wider questions that, in some cases, touch on questions concerning consciousness and metaphysics. What are the evolutionary roots of intelligence? It is sometimes forgotten that, to a considerable extent, the basis of brains is chemical (Thagard 2002). From this perspective, the origins of animal intelligence may be much deeper in their history than often imagined, while the emergence of parallel systems, notably in the plants (Trewavas 2003; Baluska et al. 2006), should hardly be a cause of surprise.

Are there limits to intelligence? In the case of hominids, Hofman (2001) argues that there is little room for expansion of the brain. This is largely a consequence of the different allometric trajectories of white and grey matter, whereby ever-increasing brain sizes lead to a disproportionate increase in white matter at the expense of grey matter, thereby jeopardizing the capacity for neuronal integration. In addition, the proliferation of modules in the more advanced mammals presents multiplying problems for their interconnectivity. With increasing size of the brain, so a larger and larger proportion of the neurons is given over to maintaining connections. Hofman (2001) is careful to point out that these estimates of size limitation of the hominid brain are extrapolations and that novel methods of integration and modularity might arise, but his conclusion is "that, as a species, *Homo sapiens* is nearly at the end of the road for brain evolution" (125). He goes on to point out the next step, if we chose to take it, involves applied technology. While the capacity in this context for superfast processing is self-evident, the likely emergence of other human capacities that ultimately we may value more highly seems distinctly less obvious.

Metaphysical Implications for Evolutionary Convergence

The evolution of intelligence must be in some way connected to consciousness. The latter topic has, of course, been the subject of intense discussion, the only result of which has been its inconclusiveness. On the assumption that it is not something that is simply emergent, let alone something to be explained away as some sort of category mistake, I would suggest that convergence points to some new avenues that might be worth investigating. One intriguing thought is to try to develop Oakes' (2007) application of adaptive zones. If, as he argues, birds need an atmosphere and cetaceans an ocean, then does not the brain require an equivalent mental environment in which to function? This is what Oakes refers to as "mental air." The consequences of this are intriguing and return us inevitably to some of the oldest and deepest philosophical and theological debates. Thus, if mind is adaptive to a real universe constrained by natural law, then the search for extraterrestrial intelligence is not an absurd gamble but as strong a prediction as finding that the long-sought-after sentient extraterrestrial is looking at us with camera eyes. Moreover, if Oakes' concept of "mental air" is valid, then this suggests that the existence of mathematics and logic are not accidents of the human condition, culturally and anthropologically constrained, but rather genuine universals.

As Oakes (2007) points out, this line of argument leads inexorably to idealism, a stance that he correctly notes is regarded with the deepest suspicion by most scientists. Equally reluctantly, scientists too often dismiss metaphysics as unprovable fairy tales. But what is the alternative? "Theories of everything" are incoherent, while any scientistic explanation of the way things are is simply ad hoc and quite unable to explain not only the emergence of complexity but, more importantly, how the integration of systems is maintained at all levels. To say that any of this is in the broadest sense "adaptive" simply begs the question of how the world came to be ordered in the first place. It is, of course, no accident that the incidental, the chance occurrence, the contingent happenstance, is so influential in our deracinated and nihilistic culture, especially as reflected in the biological sciences that have spent the last century trying to square the circle of a meaningless process, that is, evolution, leading to the appearance of a sentient species that sees meaning all around itself. The consequences of this world picture may be disastrous. For example, a much darker side to this belief system is the unrestricted extension of biotechnology. The mixture of contempt, condescension, and deliberate negligence of some exponents of these technologies is noteworthy. Nor need the opposition be simply that of ignorant luddites. If the world is an organized and coherent entity whose goodness we have lost sight of, then it is not naïve to talk of retracing our steps. From this perspective, to let the newly appointed laboratory assistant with a moral training inversely proportional to technical skill loose in this world is insane.

The reader will be well aware that, in many scientific circles, although not all, the above discussion will be regarded as at best fanciful and more probably deluded. Nevertheless, I will continue to argue that biology may be much closer to metaphysics than it often cares to acknowledge. In this context, I draw upon a particularly interesting example of mental activity, the creation of music. In a stimulating essay, Patricia Gray and colleagues (2001) point out, first, the striking convergences between human and animal music. They then innocently plant a bombshell: could it be that the similarities in these musics arise because there is a universal music, a cosmic harmony, that is "discovered" by sentient animals? In this sense, the metaphors of navigation and discovery, of exploration and maps, ring true in our description of evolution. Darwin effectively provided not only an explanation but also a compass.

It is a popular conjecture that, before humans spoke, they sang. That they might have done this for sexual selection, and here the convergent evolution of the descended larynx in a number of animals, including the red-deer (Fitch and Reby 2001), is possibly an interesting pointer, is ultimately beside the point. If ultimate realities exist, then their discovery by natural processes does not mean that the proximate mechanism is the sole function. Thus, in the case of human language, while its evolutionary antecedents are now partially understood, the fact remains that our discovery via semantics and syntax of an apparently infinite world of abstractional possibilities now separates us from other species. Is this process of linguistic discovery to find new forms of thought easy? Of course not, no more so than is the case for music. Such evolutionary developments are rare, but they are both convergent; and, as and when these biological properties emerge, they have the potential of irreversibly transforming the world.

In this context, it is worth, perhaps, recalling that the suspicion of a metaphysic in biology, with the consequent abandonment of any search for deeper meanings, has an alarming counterpart in the humanities. In this framework, where language (and music) is regarded as a mere evolutionary accident, is it so surprising that the humanities can be so readily poisoned and corrupted by the postmodernist enterprise? Suppose, however, that, if evolution is effectively the motor whereby the deeper realities of the universe may be uncovered, then it might be that an idealistic program can help to expel the corrosive relativism that attempts to etch our framework of meaning.

To conclude, science necessarily works in a naturalistic framework, but the identification of any general principle immediately begs foundational issues. The evidence for evolutionary convergence provides a counterpart to natural selection inasmuch as it starts to delineate a landscape across which the Darwinian mechanism operates. If, in addition, we can answer more coherently Schroedinger's primal question of "what is life?" then we may be a little closer to a general theory of evolution. If so, then this inevitably poses questions of metaphysics. Whether we choose to address them is another matter.

Acknowledgments

I am very grateful for the support of the John Templeton Foundation in the organization and facilitation of the meeting in the Vatican Observatory, Castel Gandolfo. I also thank in particular Mary Ann Meyers and George Coyne for their enthusiastic help. Sandra Last and Vivien Brown are warmly thanked for heroic typing, as well as handling with great efficiency organizational and editorial matters in Cambridge.

References

Amoutzias, G. D., D. L. Robertson, S. G. Oliver, and E. Bornberg-Bauer. 2004. Convergent evolution of gene networks by single-gene duplications in higher eukaryotes. *EMBO Rep.* 5: 274–79.

Axe, D. D. 2004. Estimating the prevalence of protein sequences adopting functional enzyme folds. *J. Mol. Biol.* 341: 1295–1315.

Baluska, F., S. Mancuso, and D. Volkmann, eds. 2006. *Communication in plants: Neuronal aspects of plant life.* Berlin: Springer.

Bateman, R. M., and W. A. DiMichele. 1994. Heterospory: The most iterative key innovation in the evolutionary history of the plant kingdom. *Biol. Rev.* 69: 345–417.

Beckers, G. J. L., B. S. Nelson, and R. A. Suthers. 2004. Vocal-tract filtering by lingual articulation in a parrot. *Curr. Biol.* 14: 1592–97.

Bennett, A. F. 2003. Experimental evolution and the Krogh principle: Generating biological novelty for functional and genetic analysis. *Physiol. Biochem. Zool.* 76: 1–11.

Blackburn, D. G. 1992. Convergent evolution of viviparity, matrotrophy, and specialisations for fetal nutrition in reptiles and other vertebrates. *Amer. Zool.* 32: 313–21.

Bokma, F. 2004. Evidence against universal metabolic allometry. *Funct. Ecol.* 18: 184–87.

Bond, A. B., A. C. Kamil, and R. P. Balda. 2003. Social complexity and transitive inference in corvids. *Anim. Behav.* 65: 479–87.

Bottoni, L., R. Massa, and D. L. Boero. 2003. The grey parrot (*Psittacus erithacus*) as musician: An experiment with the Temperate Scale. *Ethol. Ecol. Evoln.* 15: 133–41.

Bowman, R. 2003. Apparent cooperative hunting in Florida scrub-jays. *Wilson Bull.* 115: 197–99.

Boyce, C. K., and A. H. Knoll. 2002. Evolution of developmental potential and the multiple independent origins of leaves in Paleozoic vascular plants. *Paleobiol.* 28: 70–100.

Brown, P., T. Sutikna, M. J. Morwood, R. P. Soejono, Jatmiko, E. W. Saptomo, and R. A. Due. 2004. A new small-bodied hominin from the Late Pleistocene of Flores, Indonesia. *Nature* 431: 1055–61.

Bull, J. J., M. R. Badgett, H. A. Wichman, J. P. Huelsenbeck, D. M. Hillis, A. Gulati, C. Ho, and I. J. Molineaux. 1997. Exceptional convergent evolution in a virus. *Genetics* 147: 1497–1507.

Caporale, L. H. 2003a. Natural selection and the emergence of a mutation phenotype: An update of the evolutionary synthesis considering mechanisms that affect genome variation. *Ann. Rev. Microbiol.* 57: 467–85.

———. 2003b. Foresight in genome evolution. *Amer. Sci.* 91: 234–41.

Carosa, E., Z. Kozmik, J. E. Rall, and J. Piatigorsky. 2002. Structure and expression of the scallop Ω-crystallin gene. Evidence for convergent evolution of promoter sequences. *J. Biol. Chem.* 277: 656–64.

Chappell, J., and A. Kacelnik. 2004. Selection of tool diameter by New Caledonian crows *Corvus moneduloides. Anim. Cognit.* 7: 121–27.

Cirelli, C. 2003. Searching for sleep mutants of *Drosophila melanogaster. BioEssays* 25: 940–49.

Conant, C., and A. Wagner. 2003. Convergent evolution of gene circuits. *Nature Genetics* 34: 264–66.

Conway Morris, S. 2003. *Life's solution: Inevitable humans in a lonely universe.* Cambridge: Cambridge University Press.

Cooper, T. F., D. E. Rozen, and R. E. Lenski. 2003. Parallel changes in gene expression after 20,000 generations of evolution in *Escherichia coli. Proc. Natl. Acad. Sci. USA* 100: 1072–77.

Cresko, W. A., A. Amores, C. Wilson, J. Murphy, M. Currey, P. Phillips, M. A. Bell, C. B. Kimmel, and J. M. Postlethwait. 2004. Parallel genetic basis for repeated evolution of armor loss in Alaskan threespine stickleback populations. *Proc. Natl. Acad. Sci. USA* 101: 6050–55.

Denton, M. J., C. J. Marshall, and M. Legge. 2002. The protein folds as platonic forms: New support for the pre-Darwinian conception of evolution by natural law. *J. Theoret. Biol.* 219: 325–42.

Diamond, J., and A. B. Bond. 2003. A comparative analysis of social play in birds. *Behaviour* 140: 1091–1115.

Earl, D. J., and M. W. Deem. 2004. Evolvability is a selectable trait. *Proc. Natl. Acad. Sci. USA* 101: 11531–36.

Emery, N. J., and N. S. Clayton. 2004. The mentality of crows: Convergent evolution of intelligence in corvids and apes. *Science* 306: 1903–7.

Enggist-Dueblin, P., and U. Pfister. 2002. Cultural transmission of vocalizations in ravens, *Corvus corax. Anim. Behav.* 64: 831–41.

Farmer, C. G. 2000. Parental care: The key to understanding endothermy and other convergent features in birds and mammals. *Amer. Nat.* 155: 326–34.

———. 2003. Reproduction: The adaptive significance of endothermy. *Amer. Nat.* 162: 826–40.

Fernald, R. D. 2000. Evolution of eyes. *Curr. Opin. Neurobiol.* 10: 444–50.

Fitch, W. T., and D. Reby. 2001. The descended larynx is not uniquely human. *Proc. Roy. Soc. Lond. B* 268: 1669–75.

Flemming, A. F., and D. G. Blackburn. 2003. Evolution of placental specializations in viviparous African and South America lizards. *J. Exp. Zool. A* 299: 33–47.

Gray, P. M., B. Krause, J. Atema, R. Payne, C. Krumhansl, and L. Baptista. 2001. The music of nature and the nature of music. *Science* 291: 52–54.

Grime, J. P. 2001. *Plant strategies, vegetation processes, and ecosystem properties.* Chichester, UK: Wiley.

Hemsley, A. R., J. Lewis, and P. C. Griffiths. 2004. Soft and sticky development: Some underlying reasons for microarchitectural pattern convergence. *Rev. Palaeobot. Palynol.* 130: 105–19.

Hoekstra, H. E., and M. W. Nachman. 2003. Different genes underlie adaptive melanism in different populations of rock pocket mice. *Mol. Ecol.* 12: 1185–94.

Hofman, M. A. 2001. Brain evolution in hominids: Are we at the end of the road? In *Evo-*

lutionary anatomy of the primate cerebral cortex, ed. D. Falk and K. R. Gibson, 113–27. Cambridge: Cambridge University Press.

Horn, H. 1999. Evolutionary efficiency and success in monitors: A survey on behaviour and behavioural strategies and some comments. *Mertensiella* 11: 167–80.

Hillebrand, H. 2004. On the generality of the latitudinal diversity gradient. *Amer. Nat.* 163: 192–211.

Hunt, G. R. and R. D. Gray. 2003. Diversification and cumulative evolution in New Caledonian crow tool manufacture. *Proc. Roy. Soc. Lond. B* 270: 867–74.

———. 2004. The crafting of hook tools by wild New Caledonian crows. *Proc. Roy. Soc. Lond. B (Suppl.)* 271: S88–S90.

Huxman, T. E., M. D. Smith, P. A. Fay, et al. 2004. Convergence across biomes to a common rain-use efficiency. *Nature* 429: 651–54.

Immesberger, A., and T. Burmester. 2004. Putative phenoloxidases in the tunicate *Ciona intestinalis* and the origin of the arthropod hemocyanin superfamily. *J. Comp. Physiol. B* 174: 169–80.

Jobbagy, E. G., and R. B. Jackson. 2000. Global controls of forest line elevation in the northern and southern hemispheres. *Global Ecol. Biogeogr.* 9: 253–68.

Kaiser, W. 1988. Busy bees need rest, too. Behavioural and electromyographical sleep signs in honeybees. *J. Comp. Physiol.* 163: 565–84.

Kelly, D. A. 2002. The functional morphology of penile erection: Tissue designs for increasing and maintaining stiffness. *Integr. Comp. Biol.* 42: 216–21.

———. 2004. Turtle and mammal penis designs are anatomically convergent. *Proc. Roy. Soc. Lond. B (Suppl.)* 271: S293–S295.

Kenward, B., C. Rutz, A. A. S. Weir, J. Chappell, and A. Kacelnik. 2004. Morphology and sexual dimorphism of the New Caledonian crow *Corvus moneduloides*, with notes on its behaviour and ecology. *Ibis* 146: 652–60.

King, S., F. Beck, and U. Lüttge. 2004. On the mystery of the golden angle in phyllotaxis. *Plant Cell Environ.* 27: 685–95.

Kurtz, D. M. 1999. Oxygen-carrying proteins: Three solutions to a common problem. *Essays Biochem.* 34: 85–100.

Lahr, M. M., and R. Foley. 2004. Human evolution writ small. *Nature* 431: 1043–44.

Lanham, E. J., and C. M. Bull. 2004. Enhanced vigilance in groups of *Egernia stokesii*, a lizard with stable social aggregations. *J. Zool., Lond.* 263: 95–99.

Lefebvre, L., S. M. Reader, and D. Sol. 2004. Brains, innovations and evolution in birds and primates. *Brain Behav. Evoln.* 63: 233–46.

Levey, D. J., R. S. Duncan, and C. F. Levins. 2004. Use of dung as a tool by burrowing owls. *Nature* 431: 39.

Linden, P. F., and J. S. Turner. 2004. "Optimal" vortex rings and aquatic propulsion mechanisms. *Proc. Roy. Soc. Lond. B* 271: 647–53.

Matos, M., P. Simôes, A. Duarte, C. Rego, T. Avelar, and M. R. Rose. 2004. Convergence to a novel environment: Comparative method versus experimental evolution. *Evolution* 58: 1503–10.

Maurer, B. A., J. H. Brown, T. Dayan, et al. 2004. Similarities in body size distributions of small-bodied flying vertebrates. *Evoln. Ecol. Res.* 6: 783–97.

Morwood, M. J., R. P. Soejono, R. G. Roberts, T. Sutikna, C. S. M. Turney, K. E. Westaway, W. J. Rink, J-x. Zhao, G. D. van den Bergh, R. A. Due, D. R. Hobbs, M. W. Moore, M. I. Bird, and L. K. Fifield. 2004. Archaeology and age of new hominin from Flores in eastern Indonesia. *Nature* 431: 1087–91.

Mundy, N. I., N. S. Badcock, T. Hart, K. Scribner, K. Janssen, and N. J. Nadeau. 2004. Conserved genetic basis of a quantitative plumage trait involved in mate choice. *Science* 303: 1870–73.

Nakatsu, C. H., R. Korona, R. E. Lenski, F-J. de Bruijn, T. L. Marsh, and L. J. Forney. 1998. Parallel and divergent genotypic evolution in experimental populations of *Ralstonia* sp. *J. Bacteriol.* 180: 4325–31.

Niklas, K. J. 1991. Flexural stiffness allometries of angiosperm and fern petioles and rachises: Evidence for biomechanical convergence. *Evoln.* 45: 734–50.

Niklas, K. J., and B. J. Enquist. 2001. Invariant scaling relationships for inter-specific plant biomass production rates and body size. *Proc. Natl. Acad. Sci. USA* 98: 2922–27.

Nudds, R. L., G. K. Taylor, and A. L. R. Thomas. 2004. Tuning of Strouhal number for high propulsive efficiency accurately predicts how wingbeat frequency and stroke amplitude relate and scale with size and flight speed in birds. *Proc. Roy. Soc. Lond. B* 271: 2071–76.

Oakes, E. T. 2007. Complexity in context: The metaphysical implications of evolutionary theory. In *Fitness of the cosmos for life: Biochemistry and fine-tuning*, ed. J. D. Barrow, S. Conway Morris, S. J. Freeland, and C. L. Harper, 49–69. Cambridge: Cambridge University Press.

O'Connor, D., and R. Shine. 2003. Lizards in "nuclear families": A novel reptilian social system in *Egernia saxatilis* (Scincidae). *Mol. Ecol.* 12: 743–52.

Pack, A. A., and L. M. Herman. 2004. Bottlenosed dolphins (*Tursiops truncatus*) comprehend the referent of both static and dynamic human gazing and pointing in an object-choice task. *J. Comp. Psychol.* 118: 160–71.

Perfeito, L., L. Fernandos, C. Mota, and I. Gordo. 2007. Adaptive mutations in bacteria: High rate and small effects. *Science* 317: 813–15.

Porter, M. L., and K. A. Crandall. 2003. Lost along the way: The significance of evolution in reverse. *Trends Ecol. Evoln.* 18: 541–47.

Ramón, F., J. Hernandez-Falcon, B. Nguyen, and T. H. Bullock. 2004. Slow wave sleep in crayfish. *Proc. Natl. Acad. Sci. USA* 101: 11857–61.

Rehkämper, G., and K. Ziller. 1991. Parallel evolutions in mammalian and avian brains: Comparative cytoarchitectonic and cytochemical analysis. *Cell Tissue Res.* 263: 3–28.

Reich, P. B., D. S. Ellsworth, and M. B. Walters. 1998. Leaf structure (specific leaf area) modulates photosynthesis-nitrogen relations: Evidence from within and across species and functional groups. *Funct. Ecol.* 12: 948–58.

Reid, S. D., C. J. Herbelin, A. C. Bumbaugh, R. K. Selander, and T. S. Whittam. 2000. Parallel evolution of virulence in pathogenic *Escherichia coli*. *Nature* 406: 64–67.

Reiss, D., and L. Marino. 2001. Mirror self-recognition in the bottlenose dolphin: A case of cognitive convergence. *Proc. Natl. Acad. Sci. USA* 98: 5937–42.

Sauer, S., M. Kinkelin, E. Herrmann, and W. Kaiser. 2003. The dynamics of sleep-like behaviour in honey bees. *J. Comp. Physiol. A* 189: 599–607.

Savage, V. M., J. F. Gillooly, W. H. Woodruff, G. B. West, A. P. Allen, B. J. Enquist, and J. H. Brown. 2004. The predominance of quarter-power scaling in biology. *Funct. Ecol.* 18: 257–82.

Schweitzer, M. H., and C. L. Marshall. 2001. A molecular model for the evolution of endothermy in the theropod-bird lineage. *J. Exp. Zool.* 291: 317–38.

Sol, D., S. Timmermans, and L. Lefebvre. 2002. Behavioural flexibility and invasion success in birds. *Anim. Behav.* 63: 495–502.

Stebbins, E. C., and J. E. Galan. 2001. Structural mimicry in bacterial virulence. *Nature* 412: 701–5.

Sucena, E., I. Delon, I. Jones, F. Payre, and D. L. Stern. 2003. Regulatory evolution of *shavenbaby/ovo* underlies multiple cases of morphological parallelism. *Nature* 424: 935–38.

Thagard, P. 2002. How molecules matter to mental computation. *Phil. Sci.* 69: 429–46.

Tibbetts, E. A. 2002. Visual signals of individual identity in the wasp *Polistes fuscatus*. *Proc. R. Soc. Lond.* B 261: 1423–28.

Tocheri, M. W., C. M. Orr, S. G. Larson, T. Sutikna, Jatmiko, E. W. Saptomo, R. A. Due, T. Djubiantono, M. J. Morwood, and W. L. Jungers. 2007. The primitive wrist of *Homo floresiensis* and its implications for hominin evolution. *Science* 317: 1743–45

Trewavas, A. 2003. Aspects of plant intelligence. *Ann. Bot.* 92: 1–20.

Tripp, B. C., K. Smith, and J. G. Ferry. 2001. Carbonic anhydrase: New insights for an ancient enzyme. *J. Biol. Chem.* 276: 48615–18.

Vertes, R. P. 2004. Memory consolidation in sleep: Dream or reality. *Neuron* 44: 135–48.

Vowles, E. J., and W. Amos. 2004. Evidence for widespread convergent evolution around human microsatellites. *PLoS Biol.* 2: 1157–67.

Wagner, P. J. 2000. Exhaustion of morphologic character states among fossil taxa. *Evoln.* 54: 365–86.

Walker, M. P., and R. Stickgold. 2004. Sleep-dependent learning and memory consolidation. *Neuron* 44: 121–33.

West, G. B., J. H. Brown, and B. J. Enquist. 1999. The fourth dimension of life: Fractal geometry and allometric scaling of organisms. *Science* 284: 1677–79.

Weill, M., A. Bertomieu, C. Berticat, G. Lutfalla, V. Nègre, N. Pasteur, A. Philips, J-P. Leonetti, P. Fort, and M. Raymond. 2004. Insecticide resistance: A silent base prediction. *Curr. Biol.* 14: R552–R553.

Weinreich, D. M., N. F. Delaney, M. A. DePristo, and D. L. Hartl. 2006. Darwinian evolution can follow only very few mutational pathways to fitter proteins. *Science* 312: 111–14.

Wichman, H. A., M. R. Badgett, L. A. Scott, C. M. Bowlianne, and J. J. Bull. 1999. Different trajectories of parallel evolution during viral adaptation. *Science* 285: 422–24.

Wittenberg, J. B., M. Bolognesi, B. A. Wittenberg, and M. Guertin. 2002. Truncated hemoglobins: A new family of hemoglobins widely distributed in bacteria, unicellular eukaryotes and plants. *J. Biol. Chem.* 277: 871–74.

Wittkopp, P. J., B. L. Williams, J. E. Selegue, and S. B. Carroll. 2003. *Drosophila* pigmentation evolution: Divergent genotypes underlying convergent phentotypes. *Proc. Natl. Acad. Sci. USA* 100: 1808–13.

Wolfe, J. A. 1991. Palaeobotanical evidence for a June "impact winter" at the Cretaceous/Tertiary boundary. *Nature* 352: 420–23.

5 ASPECTS OF PLANT INTELLIGENCE

Convergence and Evolution

Anthony Trewavas

A goal for the future would be to determine the extent of knowledge the cell has of itself and how it utilises this knowledge in a thoughtful manner when challenged.

McClintock 1984

Introduction

Plant Signals and Behavior

This chapter takes as its general theme intelligence and, in particular, aspects of plant intelligence. Plants are not generally credited with intelligent behavior and the term *vegetable* is often used to describe "brain-dead" human individuals. However the acquisition of photosynthesis several eons back by the primordial eukaryotic cell obviated any need for movement to acquire external energy, unlike nonphotosynthetic animal cells. There is little doubt that the apparent paucity of movement is the main reason that plants are assumed to be "brain-dead," that is, lacking in all aspects of intelligent behavior.

Time lapse facilities, a burgeoning plant-signal transduction field, detailed analysis of the dynamic nature of the plant phenotype, and continued advances in all aspects of plant physiology, ecology, molecular biology, and cell-to-cell communication paint an entirely different picture (Trewavas 1999, 2000). Plant behavior is defined as the response to signals (Silvertown and Gordon 1988), and a plethora of external signals are sensed and acted upon by green plants. Resources (light, minerals, and water) figure strongly in a signals list that also includes numerous mechanical influences such as wind, rain, and touch; gases such as ethylene and nitric oxide; soil compaction and particle structure; and numerous biotic features, such as identity of neighbors and disturbance, among many others (Trewavas 2000; Gilroy and Trewavas 2001; Callaway et al. 2003). Changes in architecture and modifications of phenotype and physiology are used to dynamically optimize resource foraging by the individual (De Kroon and Hutchings 1995; Evans and Cain 1995; Grime et al. 1986, Grime 1994; Hutchings and De Kroon 1994; Slade and Hutchings 1987). The plant phenotype is strikingly plastic (Bradshaw 1965; Callaway et al. 2003; Jennings and Trewavas 1986; Schlichting and Pigliucci 1998; Sultan 2000). As well as improving survival, plasticity helps to deny resources to other individuals by active competition (Aphalo and Bellare 1995).

Plants use resources (i.e., food) as signals because resource availability is subject to continual spatial and temporal change. Although the shoot's physical environment changes throughout seconds to months on both a regular and chaotic basis, the individual plant also modifies its own environment by continued resource exploitation and growth. Furthermore, present signals are used to predict likely future changes in resource supply. The plant body is, then, plastically restructured to prepare for such an eventuality (if it should come) so as to maintain resource acquisition in the face of competing neighbors or environmental change. Optimizing resource acquisition is important for all plants because stored resources are a strong determinant of seed number, a crucial component of fitness. Navigating any complex environment to optimize fitness requires intelligent behavior; those best able to master their local environment are those most likely to succeed (Corning 2003).

Plants and animals differ fundamentally in the way they express behavior in response to signals. *In plants, it is phenotypic plasticity; in animals, it is movement* (Trewavas 2003). Once that behavioral difference is appreciated, the notion of plant intelligence becomes easier to appreciate. Animals efficiently forage for food using a mixture of exploratory activity followed by spatially directed movement toward detected food sources. In a precisely analogous fashion, green plants speculatively explore their local environment by occasional production of shoots and roots followed by spatially targeted shoot or root proliferation once an optimal resource supply is detected (Grime et al. 1986; De Kroon and Hutchings 1995; Hutchings and De Kroon 1994). Many plants are very efficient in targeting leaves and roots to areas of high resource supply, enabled by a remarkable degree of morphological plasticity. It is in foraging for food that animal intelligence becomes a premium, and it is in plant foraging that plant intelligence comes to the fore.

Information processing in plants

The picture of plants that has emerged in the last decade is of a complex and sensitive information-processing organism (Aphalo and Bellare 1995). But this behavior is largely invisible to us because the different time scale makes observation difficult without patient, careful, and detailed measurement. Furthermore, the difficulties of observing the plant below ground still obstruct the development of a proper perspective over plant behavior, although many recent studies are indicating some remarkably unusual capabilities (e.g., Falik et al. 2003; Gruntman and Novoplansky 2004; Schenk et al. 1999). However, such behavior did not escape Darwin, as we shall see later, who observed and wrote down his observations in great detail. Time lapse frequently provides a much better perspective than that available to Darwin (e.g., Attenborough 1995). It is crucial to observe plants under conditions that mimic those in the wild. If we consider a domesticated species such as *Brassica oleracea*, which includes kale, broccoli, cauliflower, brussel sprouts, and cabbage (Pink and Puddephat 1999), we encounter a plant breeding success story and an indication of the extent of phenotypic plasticity, but such plants do not exist in the wild. Domesticated species have had their morphology and behavior restricted for our benefit, but none survives in fallow fields where they must compete with behaviorally adept wild species. Greenhouses, window sills, and laboratory-grown plants lack competition as well as environmental chaos and can seriously mislead. McClintock (1984), whose quotation above actually comes from her Nobel Prize acceptance speech, was a plant biologist credited with the discovery of transposons. Like many plant biologists who have studied their chosen organism for many years, she came away with a deep respect for the skill with which plants surmount their environmental problems by adaptation. To paraphrase McClintock's rather poetic language, "a goal for the future would be to determine the structure of the current integrated molecular network of the cell (organism) resulting from development and environmental experience (memory) and how that network acts to intelligently generate successful adaptive responses when signalled." As we acquire more knowledge about all sorts of behavioral characteristics of living organisms, not only are previous assessments of intelligence and behavior shown to be wrong, but the expanding view enlarges our perspective of life itself.

The Biological Meaning of Intelligence

Background

Dictionaries define intelligence either as reasoning or sentience. Many psychologists consider cognitive abilities the crucial index of human intelligence but will occasionally accept adaptation as indicative of intelligence in other organisms. But human intelligence can, in turn, be considered as merely a complex version of human adaptation. It is commonly assumed, however, that only higher mammals or just human beings are really capable of intelligent behavior. Occasionally, cetaceans, even some birds, particularly corvids, may be allowed into the intelligent club. Such restrictions may reflect the emphasis given by psychologists to the measurement of human intelligence, although a psychology encyclopedia published in 1942 contained a chapter on plant behavior (Warden, Jenkins, and Warner 1942). Even if this restriction is relaxed, a further limitation is imposed with the assumption that only organisms with brains can be intelligent. This very crude form of brain chauvinism (Vertosick 2002) is both anthropocentric and credits nerve cells with some sort of vitalistic quality (Schull 1990).

The word *intelligence* derives from the Latin *inter-legere*, literally to "choose between," and a critical trap to avoid in this area is the easily discredited notion of "subjective intelligence" (Warwick 2001); that is, defining intelligence only within human capabilities. In assessing the intelligence of other species, Warwick states the following:

> Comparisons are (usually) made between characteristics that humans regard as important. Such a stance is, of course, biased and subjective . . . in terms of the groups by which it is being viewed.
>
> Warwick 2001, 160

> The success of a species depends on it performing well in its own particular environment and intelligence plays a critical part in this success.
>
> Warwick 2001, 9

> When we compare the important aspects of intelligence between species, it is those which can allow one species to dominate and exert power over other species that are of prime importance.
>
> Warwick 2001, 213

Stenhouse (1974) in discussing the evolution of intelligence described it as "adaptively variable behaviour during the lifetime of the individual" in an attempt to discriminate between innate responses and intelligence. The emphasis on the individual is important and accords with common views on natural selection that equally emphasize individual variation. However, as will be seen later, it is more likely that innate behavior arose from learned, that is, intelligent, behavior in the first place, potentially by genetic assimilation. Many biologists have used the broader description of intelligence as choice, and their recognition of the kinds of observable biological intelligence is indicated in the sections below.

Intelligence consists of two distinguishable components. There is the organism capable of intelligent responses, and there is the environmental problem that necessitates the application of intelligence for its potential solution. Intelligent behavior will not emerge or be detected without the right circumstances (such as finding food) to elicit it.

1. Species Intelligence

"Plant and animal species are information processing entities of such complexity, integration and adaptive competence that it may be scientifically fruitful to regard them as intelligent" (Schull 1990, 63). In a detailed and complex article critically discussed in detail by his psychological peers,

Schull (1990) has cogently argued the case for individual species to be regarded as intelligent. The nub of his arguments hinges around the very numerous analogies between species behavior and learning in higher animals. Schull argues for strong parallels between Wright's evolutionary adaptive landscape (Wright 1982) and the Darwinian mechanisms used to select successful neural pathways that underpin learned behavior in the developing brain. Even Lamarck (1809) regarded the evolution of a new species as like birth (or asexual reproduction by fission) and its subsequent extinction as like death. Gould (2002), in a very lengthy and persuasive dissertation based on systems approaches, argues the case for evolution to occur at the species level as well as that of the individual organism. The case for a species to be regarded as an individual entity, as required by Schull, is thus strengthened.

The notion of species as an identifiable integrated entity also follows from simple hierarchical analysis of biology that places individual species as one level between molecules, cells, and individuals up to ecosystems (Trewavas 1998). Species are composed of numerous individuals constructed into a network through their ability to reproduce and compete with each other. But, importantly, the ability of groups of individuals to cluster into demes with varying degrees of selection success provides necessary variety in the strength of the connections within the network (Wright 1982).

Schull (1990) convincingly draws parallels between genetic assimilation mechanisms (Waddington 1957; Schmalhausen 1949) and foresight, the ability that allows organisms to come up with a behavioral solution to an environmental problem with minimal trial and error. Improved behavioral modification enables the subsequent selection of genes and gene combinations in demes that allow the strategy to develop with greater rapidity, higher probability, or lower cost (Bateson 1963). Consequently, evolution becomes much faster than mechanisms that require selection of random gene combinations, just as foresight reduces the time required for successful behavior. Schmalhausen (1949) uses the familiar example of numerous water plants such as *Saggitaria* and *Myriophyllum*, which exhibit two completely different phenotypes according to growth in water or on land, as examples of anticipation of future environmental variation and genetic assimilation. Other closely related water plant species may exhibit just the one kind of phenotype regardless of environment (Henslow 1908).

2. Bacterial Intelligence

The case for bacterial intelligence is based on three observations of bacterial behavior (Miller and Bassler 2001; Park et al. 2003a,b; Vertosick 2002).

A. Bacteria acquire enormous pieces of genetic information via sexual exchange, viral-mediated transfer, and transduction and transformation between species. This alone accounts for their skill in dealing with antibiotics and represents an informational network that can extrapolate from the past and react to novel situations using that experience as a guide. The strength of connections within the network is determined by the probability of information transfer among or between species.

B. Intercellular communication (quorum sensing) is now known to occur in the formation of biofilms, communities constructed from a network of millions of cooperating individuals. Biofilms form via chemotaxis, with complex wave patterns generated that can be observed using microfabricated mazes (Park et al. 2003 a,b). Emergent properties (such as luminescence or virulence) develop when the communities reach certain critical sizes as a result of auto-inducing signals surpassing threshold concentrations (Miller and Bassler 2001). It is increasingly thought that biofilms are the natural bacterial state and that individuals growing separately in solution are experimental artifacts (Bassler 2002).

C. Mutation rate increases up to a one-thousand-fold higher in response to stress (Cairns et al. 1988).

Bacterial memory resides in the huge numbers of species of bacteria that can transmit genetic information to each other. The population learns by the production of colony individuals with new capabilities that gain advantage and replicate at the expense of others. In this case, learning results from "natural" selection of the most efficient individual in a network of different individuals. (Analogously, a consistent theme in all hypotheses of learning processes in the brain is the Darwinian selection of the most efficient neural pathway). Bacterial colonies should not be regarded as random collections of unintelligent individuals but are instead organized into nonlinearly interacting societies of genetically variable individuals connected by a constant interplay of hormonal (auto-inducing) signals and genetic information. Even in the *lac* operon response, individual cell variation accounts for an apparent dose response of induction against lactose concentration (Novick and Weiner 1957).

3. Protozoan Intelligence

The slime mold *Physarum* can solve a simple problem of the shortest distance between two points when faced with a maze offering variable length solutions to a food source at the end. "This remarkable process of cellular computation implies that cellular materials can show a primitive intelligence" (Nakagaki et al. 2000, 470). Amoebae will prey on *Tetrahymena* but avoid *Copromonas*. If given the choice, *Paramecium* will feed on small ciliates but not on bacteria (Corning 2003).

4. Intelligent Genomes

Thaler (1994) in the article "Evolution of Genetic Intelligence" discusses the controversial results produced by Cairns et al. (1988) and Hall (1992). He outlines the case for increased stress conditions generating mutation, pointing to a feedback loop between the environment and the generators of genetic diversity such as mobile elements like transposons and genes that respond to the stress (see also McClintock 1984). In plants, it is known that many different stresses cause transposon movements (Kumar and Bennetzen 1999) and that bacterial infection or UV stress causes chromosome rearrangements (Kovalchuk et al. 2003; Ries et al. 2000). Durrant (1962, 1981) observed that imbalance of mineral treatment in plants caused adaptive heritable changes lasting some ten to twelve generations that slowly revert to the initial parental type. Similar observations suggest that long-term heritable changes in both flower structure and growth of many generations of siblings can result from an initial environmental impact on the parent (Highkin 1958, Hill 1965). Parental environmental conditions are known to influence the disease resistance behavior of subsequent seedlings for several generations (Agrawal et al. 1999), indicating that Cairns-type mechanisms may be common in plants but remain largely unrecognized.

These observations on plants can be made because, unlike many animals, plants grow and develop throughout their life cycle. Embryogenesis continues throughout the life cycle, and the embryogenic meristems eventually form flowers. Environmental history can, thus, pass directly into reproduction. The Weismann proscription that the environment does not directly affect animal inheritance, because sexual cells are protected from environmental variation, is inapplicable to plants, strengthening the likelihood of neo-Lamarckian inheritance in plant evolution. This important difference between plants and animals was recognized very early on (Henslow 1908).

5. Immune Intelligence

The workings of immune systems have often been described as similar to animal cognition (Coutinho 2002; De Castro and Timmis 2002). Thus, the system learns to produce specific antibodies to specific antigens, and it discriminates between self and nonself and maintains a memory of past experiences that it can access. In metaphorical terms, it has been described as the "molecular consciousness of the body"; in evolutionary terms, the immune system is able to foresee the evolution of disease microbes and, thus, expresses foresight (Vertosick 2002).

The immune system learns patterns of shape. When presented with a new antigen, successful antibodies modeling that shape and binding with strong affinity to it are synthesized within several weeks. The immune system is diffuse in the body and involves millions of T and B lymphocytes. Gene shuffling in the variable region of single antibody genes is performed in response to the presence of a new antigen until new antibodies are produced by individual lymphocytes. By a process akin to natural selection, those antibodies binding most strongly are eventually selected, and the individual lymphocyte, in which they occur, undergoes massive replication. Successful immune learning results from trial-and-error Darwinian competition. Note the similarity to learning in bacteria and parallels with neural network learning by selection of the successful neural pathways (Vertosick and Kelly 1991). It is thought that interactions occur between large numbers of lymphocytes to generate the most-fit solution (Vertosick 2002). Thus, again, a network of interacting individuals is present in which the strength of particular network connections can be altered. Immune intelligence is now part of artificial intelligence studies (De Castro and Timmis 2002).

6. Swarm Intelligence

A swarm is defined as a set of mobile agents that are liable to communicate directly or indirectly with each other by acting upon their local environment and that collectively carry out distributed problem solving (Bonabeau and Meyer 2001; Bonabeau and Theraulaz 2000; Bonabeau et al. 2000). Swarm intelligence is exemplified by the social insects and was first recognized in termites and later bee and ant colonies and is now much studied because of the complexity of problems that can be solved by relatively simple rules of interaction between members of the swarm. Each individual has no fundamental knowledge of the final outcome. But simple behavioral rules can construct immense nest structures (termites), minimal paths to food and space estimation (ants), or, by adaptive changes in overall hive behavior, harvest different kinds of food by communicative behavior of workers on the dance floor (bees) (Franks et al. 2003; Seeley 1995). The swarm represents an interactive network constructed from the individuals of the nest, and different behaviors can be elicited by changing the strength of interactions between the individuals to guide them into new activities. "Indeed it is not too much to say that that a bee colony is capable of cognition in much the same sense that a human being is. The colony gathers and continually updates diverse information about its surroundings, combines this with information about its internal state and makes decisions that reconcile its well being with its environment" (Seeley and Leven 1987, 41). Green plants routinely perform the same detailed environmental perception, make meaningful assessments, and construct adaptive responses. Bell (1984) pointed to similarities in structure between plant branching morphologies and the foraging system of ants.

7. Metabolic Intelligence

The cell is composed of sets of interlocking metabolic pathways, complex signal transduction, and protein networks (Gavin et al. 2002; Maslov and Sneppen 2002; Perkel 2004; Ravasz et al. 2002) Enzymes are connected together through substrates and products, through allosteric effectors or inhibitors; other proteins undergo direct interaction through signalling pathways, cytoskeletal rearrangements, and control of gene expression. About one thousand different protein kinases (numbers are identical in both plants and animals) interlink these together with organelles into a composite collective we call the cell. Metabolic control theory has indicated that control is inevitably shared throughout the metabolic network and that many enzymatic processes behave like Boolean logic computer gates (e.g., Arkin and Ross 1994, Bray 1995, Okamoto et al. 1987). Metaphorically, the interaction of connection strengths and fluxes can be envisaged as a landscape with hills and valleys, much as Wright (1982) used a similar picture with the adaptive landscape. Again, what underpins the concept of metabolic intelligence (Vertosick 2002) is a complex network in which the strengths of connections can be modified usually by protein modification through phosphorylation or by synthesis of critical enzymes.

8. Animal Intelligence

Nerve cells form the basis of brain intelligence. Nerve cells evolved initially to speed up the connections between sensory system and responsive tissues by using action potentials to transmit information down the length of the dendrite. Research on the snail *Aplysia* indicates that learning involves the formation of new dendrites and that memory of that learned response persists as long as the dendrite is extant (Kandel 2001). When the dendrite disappears, the specific memory is lost. Each nerve cell may have many dendrites enabling many different connections to be constructed. In a complex brain, learning and memory can be expected to operate via many connecting pathways of information flow. Interactions between even larger numbers of nerve cells give rise to emergent properties in more advanced organisms of sentience and reasoning.

Thus, intelligent behavior by organisms with brains relies on a complex cellular network in which the strengths of connections can be changed by forming new dendrites or by strengthening the connections through preexisting dendrites. Learning involves Darwinian-like mechanisms in which information pathways are selected for their success in approaching the desired behavior. Edelman (1993) has also described a Darwinian selection mechanism in the embryonic brain to explain how different neural anatomies can give rise to similar behavioral traits. Although it used to be thought that electrical transmission was the critical factor in brain functioning, it is now evident that virtually all transmission of information between nerve cells is chemical. At least one hundred chemicals are involved (Greengard 2001).

How Biologists Describe Intelligence

From the discussion above, several descriptions of intelligent behavior emerge. These are:

A. to store past experiences and to use that acquired knowledge to solve future problems;

B. information processing, choice, leading to assessment and adaptive responses;

C. adaptively variable behavior during the lifetime of the individual to distinguish innate from learnt responses (Stenhouse 1974).

All such descriptions require the following:

A. detailed environmental perception resulting from signals;
B. meaningful interpretation of such signals;
C. adaptive responses as a consequence of interpretation.

Apart from the higher animals that use the centralized activity of the brain to process information and in which classical intelligence is located, *all other biological systems possess a decentralized intelligence that is a consequence of behavior by the whole system*. All forms of described intelligence above involve a network of interacting constituents of varying degrees of complexity, whether it be molecules, cells, or individual organisms, *through which information flows*. The common important property that makes the network intelligent is that the *connection strengths in the network can be varied, thus enabling information flow to be directed* into different channels as required. Different signals can be directed to produce different responses, but cohesion between different information pathways will result in synergistic, cooperative, or competitive effects between numerous signals. *The simplest forms of memory represent semipermanent alterations in the speed or the specific channel of information flow* from particular signals to their response implementor.

Plant Intelligence

Historical Recognition of Plant Behavior

"It is hardly an exaggeration to say that the tip of the root acts like the brain of one of the lower animals" (Darwin 1882, 573). "In several respects light seems to act on plants in nearly the same manner as it does on animals by means of the nervous system" (Darwin 1882, 566). "I have repeatedly had cause to refer to certain resemblances between the phenomena of irritability in the vegetable kingdom and those of the animal body (Von Sachs 1879, translated 1887, 500). "*Dass sich die lebende Pflanzensubstanz derart innerlich differenzirt dass einzelne theile mit specifischen energein aus gerustet sind, ahnlich wie die verschiedenen Sinnesnerven des Thiere* (Von Sachs 1879, quoted in Darwin 1882, 571).

Sachs and Darwin are recognizably the preeminent botanists of the nineteenth century, and both spent decades in detailed observation and experimental investigation of plant behavior—that is, the response to signals. Although it was Darwin who indicated that natural selection involved overproduction of progeny, competition for resources and "differential survival of those best able to master their environment," this active competitive element is missing from most present-day descriptions of plant behavior. The failure to recognize the dynamic of intelligent behavior required from successful wild plants results from the very limited experience most scientists have of plant behavior.

Distinguishing the Passive and the Active View of Plant Behavior

Two perceptions of plant growth and behavior need to be distinguished. A common *passive* view is that plants grow according to a predetermined genetic program with rates determined merely by provided resources (food). Occasional environmental stresses restrict growth rates. This limited view is, at best, applicable to plants in extreme conditions such as deserts or polar extremes, where plant neighbor competition is minimal or absent and the environment either unvarying or consistently varying but often powerfully stressful. But the primary reason that the passive view is common is that experience of most biologists to plants is limited to crop or horticultural plants growing under ideal growth conditions in greenhouses, fields, and, perhaps, window sills.

The *active* view of plant behavior is in complete contrast (e.g., Aphalo and Ballare 1995; Ballare 1994; Bazzaz 1996; Grime 1994; Kuppers 1994). For plants facing competition from neighbors and from other organisms in a variable abiotic environment, intelligent adaptive behavior is a necessity, not a luxury. Plants actively forage for their resources (nutrients, light, water, i.e., food) in their local environment. Changes in resource supply act as signals to reconstruct plastically the phenotype and maximize resource exploitation. At the same time, phenotypic proliferation is used actively to deny resources to competitors within the vicinity. The active view sees genes as merely specifying the limits of phenotypic change, not acting as rigid determinants of development (Sultan 2000). Resources are strongly contested in the wild and are often scarce, and plants possess numerous strategies to compete effectively. The active view is based strongly on Darwinian principles that see fierce competition as underpinning evolution and community structure (Goldberg and Barton 1992) and the need for efficient optimization of resource acquisition and exploitation.

In the few gaps available for any growing wild plant in a canopy, resources are presented as gradients in light, minerals, and available water and are effectively pulsatile in their availability (Grime 1994, Kuppers 1994, Robertson and Gross 1994). Exploratory speculative growth is common and involves minimum investment of growth resources using a morphology that ensures maximum elongation with minimum width (De Kroon and Hutchings 1995; Slade and Hutchings 1987). Shoot tissues will grow along light gradients, proliferating and thickening when light intensity increases (Gersani and Sachs 1992; Harper 1977). Those in canopy gaps experience enormous variation in shape, direction, and intensity of light that must be efficiently and plastically filled (Bazzaz 1996). Roots grow along mineral and humidity gradients, proliferating in rich resource pockets and up-regulate ion uptake mechanisms to sequester minerals and water at speed (Aphalo and Ballare 1995; Callaway et al. 2003). The resource world presented to any wild plant is a striking mosaic (Bell and Lechowicz 1994).

To an extent, plants choose the environment in which they grow (Bazzaz 1991; Salzmann 1985; Salzmann and Parker 1985). Assessment is made of the best places for exploitation, after analysis of all information; shoot branches and roots are then directed into those environmental regions

(Bazzaz 1996; Henriksson 2001; Jones and Harper 1987). Environmental information is, thus, continuously conveyed to the molecular mechanisms that manipulate morphology and physiology (Aphalo and Ballare 1995; Evans and Cain 1995; Gersani and Sachs 1992). The stilt palm, which "walks" by differential growth of prop roots out of shade into sunlight, is the most dramatic example (Trewavas 2003). Decision making about phenotypic change involves in some way the whole plant and is, thus, decentralized (Hartnett and Bazzaz 1983; Kuppers 1994; Novoplansky et al. 1989; Trewavas 2003, 2004.)

Social Organization of Plants Distinguishes Them from Animals

Plants are both social and modular organisms. Higher plants are constructed from numerous repetitions of foraging organs like leaves (buds, flowers, fruits, seeds) and branch roots (Harper 1977; Sachs et al. 1993) that form an interactive network. The numbers of such modules can vary by many millions between individuals. The common overall morphology is generally a branched structure (derived by tip growth) recognized to be the most efficient way for sessile organisms (including some bryozoans; see McKinney and McGhee 2003; McGhee and McKinney 2000) to occupy local space. In contrast, most animals are unitary organisms in which the need for movement to find food and mates necessitates a morphology constructed by a predefined, tightly regulated and replicable genetic program inside a uterus or egg. Plants are grazed continuously by animals. In that case, a decentralized morphology enables survival, whereas one with complex differentiation of function would be vulnerable to even slight damage. However, because there are numerous repetitions of growing points, behavioral integration is less easily recognized than in animals.

Information Processing Systems in Plants

To any plant, the environment is extremely complex, and plants are sensitive not only to fine changes in environmental variables but also integrate their signal information into phenotypic change. Any of the twenty or more environmental variables that plants perceive can change independently of the others. The intensity, the length of the signal, and particularly its direction are all distinguished, and the latter enables image construction (Gilroy and Trewavas 2001). Plants are separately sensitive to UV, blue, green, red, and far-red light, for example (Ballare 1994). The adaptive response varies according to the dominant wavelength or to a particular combination that changes throughout the day from seconds to hours to months (Pearcy et al. 1994). Plants are sensitive to at least seven differing degrees of water starvation (Hsaio et al. 1976), and these are distinguished by discrete physiological and phenotypic responses. Furthermore, there is strong evidence that plants are territorial and that space itself is a powerful signal altering the phenotype (Schenk et al. 1999). When individual plants are grown with equal levels of light, minerals, and water, those with greater soil space grow substantially greater. How space is perceived is not understood, but a mechanism in which branch roots of the same plant recognize each other and spatially separate themselves as far as possible may be the key.

This degree of discrimination indicates that the local environment to a plant is a far more complex affair than a simple list of light, minerals, and water implies. As plants grow, the wavelengths of perceived light changes and growing neighbors change it again, water gradients are modified on almost a minute-to-minute basis, temperature varies according to the distance from soil, and the concentration of modifying gases likewise varies as soil activities change and roots exploit local mineral patches. Wind becomes an increasingly important component of abiotic signals. And any of these environmental factors change on a daily basis and throughout the day, with weather complicating the issue further. Biotic information is also processed and results in profound phenotypic changes. These biotic signals include the presence, absence, and identity of neighbors, mutualistic interactions (particularly with fungi), herbivory, parasitism, disturbance and, most particularly, competition from other plants (Callaway et al. 2003). Each abiotic factor can vary independently.

But it would be misleading to talk about individual signals because plants sense the totality of their environment with the response to an assessed change in any one signal synergistically modified by all the others (Corning 2003; Bazzaz 1996; Trewavas 2000). The assessment must result from an integration of differing signal transduction pathways inside cells, resulting in a very complex mixture of communicating signals moving throughout the plant individual.

Receptors for some of these signals have been identified: there are at least eight different light receptors, for example, and probably more yet to be discovered (Trewavas 2000). Water (osmotic) receptors have been identified, and receptors to some growth factors have been described. Others remain yet to be characterized.

Something of the mechanisms used to process plant signals have emerged from a burgeoning field of investigation called signal transduction. Signal transduction networks are constructed in individual cells and are composed of a densely connected molecular network that has as its basis about one thousand plant protein kinases (enzymes capable of modifying the activity of many thousands of proteins by phosphorylation), numerous second messengers and many other proteins that temporarily form signal clusters attached to membranes. Connection strength through the network and thus information flow is known to be altered by phosphorylation and by synthesis of critical proteins in the network (Gilroy and Trewavas 2001; Trewavas 1999, 2000).

Competitive Foraging for Light Resources Elicits Intelligent Behavior

The shoot phenotype is generally constructed to minimize self-shading (Ackerly and Bazzaz 1995; Honda and Fisher 1978; Yamada et al. 2000). Leaves are produced at intervals from the shoot meristem. The leaf stalk (petiole) is sensitive to reflected/transmitted blue, red, and far-red radiation from other leaves, enabling plastic changes in length and direction of growth to adjust light collection (De Kroon and Hutchings 1995; Yamada et al. 2000). The pulvinus (connection of stalk to leaf) rotates the leaf lamina to face the primary direction of photosynthetically active radiation (Muth and Bazzaz 2002a). In direct competition with other plants, leaves will be inserted on top of competitors. In turn, the competitor will redirect branch and leaf growth or increase stem growth to avoid shading as far as possible.

Specific changes in light wavelength balance are responsible for changing leaf lamina direction. When growing in canopy gaps and, thus, under light competition, the orientation of leaves and the polarized direction of the branches that support them align with the primary polarized orientation of diffuse radiation (Ackerley and Bazzaz 1995). Some plants continually rotate the plane of the lamina to follow the sun's movement through the day (heliotropism). When growing in dense woodland, branches are projected into the space with strongest light intensity, and others unfavorably positioned will receive little in the way of root resource or will be closed down completely (Franco 1986; Jones and Harper 1987; Henriksson 2001; Muth and Bazzaz 2002b, 2003). Thus, the phenotype is continually adjusted to optimize light collection within the specific environmental constraints.

If the incident light is very intense, leaves may close on each other, or the lamina drops vertically to reduce light exposure. The chloroplasts are moved to the sides of cells (an actin-myosin dependent process) to reduce photo-oxidative damage. The leaf phenotype of new leaves changes as continued shoot growth places branches that were initially in direct sunlight but now are shaded (Sinnott 1960). In deep shade, leaves are abscised after removal of resources (Addicott 1982). These data emphasize that optimal foraging for light results from plasticity in the decisions about growth of branches and leaf development.

That there is vigorous competition for light is indicated by the numerous plant strategies developed to help mitigate competition, such as climbing plants and trees. If carbon resources are used either to increase root growth or to increase stem height to gain better light access over competitors, there are fewer internal resources available to provision seeds. Consequently, the seed numbers decline, and fitness is reduced. Some plants that grow in dense clumps have a specified height at which growth ceases and flowering commences. Givnish (1982) regards such behavior as altruistic since it minimizes shading by competitors of the same species. But application of game theory indicates that plants that have a strategy to maximize canopy carbon gain (that is, gain by all the members of the individual species) are simply out-competed by other species that maximize the carbon gain of the individual plant (Schieving and Poorter 1999).

However, climbing plants minimize the carbon resource required for stem strengthening by using other species as supports, thus increasing the carbon resources available for increasing seed number production. But the downside is that growth is dependent on an established canopy, and initial growth must take place under this canopy when the young plant is at its most vulnerable. The climbing solution has been solved in many different but convergent ways, using tactile petioles, leaf tendrils, stem tendrils, excitable coiling stems, and even adhesive roots (Darwin 1891). Tendrils can definitely sense plant supports within their vicinity (probably through reflected far-red radiation) and direct their movements towards them (Baillaud 1962). Tendrils can also unwind if the environmental conditions change, indicating that important "climbing" decisions can be reversed if necessary (Darwin 1891; Von Sachs 1879).

The woody plant (tree and shrub) program is the other major strategy to reduce competition for light. All the early photosynthetically gained resources are incorporated into strengthening tissues and into increased height eventually to overgrow other competitors. Consequently, long juvenile periods with no seed production and perennial life cycles are the necessary accompaniments. Juvenile periods in trees can last twenty-five to thirty years.

Foraging for Mineral Resources: Intelligent Construction of Optimal Root Networks

Mineral resources are not uniformly distributed in the soil, and rich mineral pockets are quickly exploited by local branch root proliferation (Aphalo and Ballare 1995; Drew and Saker 1975; Granato and Raper 1989; Jackson and Caldwell 1989). Exploitation shells develop rapidly in such soil regions, necessitating further phenotypic exploration of soil through branch root growth. Such observations support a pulse-patch model of root behavior (Grime 1994).

In soil with numerous individuals, there is also strong spatial segregation between the separate root systems. Competitive roots of different individuals, growing within the vicinity of each other, avoid direct contact and can cease growth if contact is forced (Callaway et al. 2003, Mahall and Callaway 1992). While architectural constraints, allelopathic chemicals and plastic responses to competition account for some of this spatial segregation, there is strong evidence that plants actively compete for space itself and are territorial, vigorously occupying local space to deny it to others (McConnaughay and Bazzaz 1991, 1992; Schenk et al. 1999). Root systems of individual plants minimize competition from their own roots, just as they minimize shading of their own leaves.

Using single plants with roots split between two pots, it has been shown that new root tissue is increasingly directed into unoccupied soil if the other pot contains increasing numbers of competitive individuals of the same species (Gersani et al. 1998). Competitors of the same species are, thus, sensed and avoided if possible. Gersani et al. (2001) used simple game theory to examine the behavior of split clones of individuals that "owned" their soil compared to those that "shared" an equivalent volume of soil and resources with another member of the same species. When forced to share soil competitively, individual plants substantially increased root proliferation around themselves to defend the mineral and water resources in their own space. In shared soil, the polarity of root growth was directed toward other competitive root systems (Gersani et al. 2001; Gruntman and Novoplansky 2004; Maina et al. 2002). However, increased root proliferation again induces a cost: that is, a reduction in subsequent seed number and, thus, fitness. The fitness-maximizing strategy of the individual plant is, thus, to sacrifice collective yield in a quest to steal nutrients from its neighbor. The implications are striking, indicating that plants are able to assess and respond to local opportunities in a manner that maximizes the good of the whole plant!

The experiments above indicate that plants can recognize themselves and recognize other individuals of the same species as competitive aliens. By dividing individual plants into separate clones, Gruntman and Novoplansky (2004) examined whether separated clones still recognized each other as self (from the same individual) or nonself, by competitively increasing root proliferation. They showed that each separated clone comes to regard other daughters from this clone as aliens within a few weeks of separated growth. Thus, individual plants are able to distinguish self from nonself and adjust their phenotype accordingly. Such observations have been confirmed in other plants (Holzapfel and Alpert 2003). Probably, the same is also true of the shoot phenotype (Schieving and Poorter 1999). Since individual species contain millions of individuals that are all regarded by any one individual as aliens, this indicates the presence of a self-recognition system of enormous complexity of which we presently have little understanding. Similar complex, self-recognition systems may be present in lower animals like sea anemones (Hart and Groseburg 1999; Groseburg and Hart 2000).

Root proliferation by the individual plant is also greatly increased if water or minerals become scarce. Resources are directed away from shoot growth and invested into increasing exploration of the soil. If carbon resources are scarce, stems become elongated and thinner to gain maximum height with the limited resources available (Bloom et al. 1985). These are decisions made by the whole plant leading to enhanced proliferation of the tissues required to recover some of the scarce materials necessary for balanced growth. The decision-making process is not yet understood.

Sophisticated Cost/Benefit Assessment Underpins Plant Resource Acquisition

It is quite clear that branches on most young plants do not grow equally. How the decisions are made to ensure that branches, best placed to seques-ter light, receive most root resources to maximize their growth is not under-stood, although competition between branches in different light environ-ments appears to operate (Aarsen 1995; Henrikkson 2001; Honkanen and Hanioja 1994). Fairly simple experiments using plants with two equal shoots but placed in differing light conditions indicate that the shoot with enhanced light conditions does grow more quickly than the other (Novo-plansky et al. 1989). If the shoot in weaker light is now placed in darkness, it eventually dies as the vascular system provisioning the shoot with root resources is sealed. But it is not just a simple assessment of light resources. A temporary mechanical constraint on a branch growing optimally can completely reverse the order of branch growth (Novoplansky 2003). It seems to be aspects of development that are crucially assessed (Novoplan-sky 1996), and the vigor of growth determines, in turn, the distribution of resources from the root, much as capital is cannily allocated to shares in companies that are expected to grow more rapidly (Novoplansky 2003). What is perceived, then, is not only the rate of growth but the anticipated future rate of growth. Since acquisition of root resources is dependent on the vascular connections, the activities of the cambium, the meristem that generates vascular cells, may be the central arbiter of resource distribu-tion and shoot phenotype construction (Sachs et al. 1993). Feed-forward and feed-back mechanisms must operate here because the assessment and decision clearly involve information from the whole plant.

Investment of growth resources only follows, however, after some sophis-ticated decision making that is most clearly observed in dodder (*Cuscuta* species), a parasitical nonphotosynthetic plant. Dodder coils around suit-able hosts within a few hours forming haustoria (suckers) that penetrate the host vascular system and sequester food after four to five days (Kelly 1990). Coiling takes several days to complete but finishes well before the acquisition of resources from the host. There is a very clear separation of resource investment to parasitize (estimated as coil length) from the subsequent energy gained from the host (which can be measured as biomass gained after several weeks). Experiments show that the decision to parasitize or not is made within a few hours of contact with the host (Kelly 1992). Commonly, dodders reject half of the offered suitable hosts, although the rate of rejection can be diminished (but not eliminated) by ensuring that the host has an abundance of nitrogenous materials. It is critical to note that dodder does not parasitize itself, indicating self-recognition.

Efficient animal foraging is characterized by a simple model developed by Charnov (1976). That is, animals normally expend least energy investment (time to find resource) for maximal energy gain. Using a number of hosts, Kelly (1990) showed that dodder foraging for resources complies with this simple Charnov foraging model. At the earliest stage of contact to the host, then, not only is a decision made by dodder whether to parasitize within a few hours, but the number of coils around the host (energy invested) is determined by this early assessment of the potential future return of resources from the host. Other papers on nonparasitical plants have confirmed the relationship of "investment" against "return," indicating sophisticated cost/benefit assessments are made in branch and root growth (Gleeson and Fry 1997; Wijesinghe and Hutchings 1999).

Plant Foresight: Predicting the Future.

Light reflected from vegetation is richer in far-red wavelengths compared to red. Plants use that information along with its direction to predict not actual shade but to foresee the likelihood of shading at some stage in the future from a competitor (Aphalo and Ballare 1995; Ballare 1994, 1999; Novoplansky et al. 1990). When a change in the balance of red to far-red radiation is perceived, an integrated adaptive response in phenotype structure results. New branches grow away from the putative competitor, stem growth is increased; the rate of branching diminishes, and such branches assume a more vertical direction; leaf area increases in anticipation of reduced incident flux; and the number of layers of leaf cells containing chlorophyll diminishes.

Foresight of future water availability also institutes characteristic morphological changes in anticipation and preparation. A single water starvation episode can engender many morphological changes in new leaf structure, reducing transpiration surface and stomatal density and increasing waxy layer and cell wall thickness to resist cell deformation and hairiness (Hsaio et al. 1976). Physiologically, osmotic adjustment is made to retain water by increasing osmolytes. Abscission of old leaves is accelerated to bring the plant back into balance with the anticipated diminished supply. These changes in leaf structure, along with enhanced changes in root structure and development, anticipate future water deficits for growth.

When single, young trees were provided with water only once in their first year, growth was sporadic throughout the growing season. In subsequent years, these plants gradually learned to predict the once-a-year water supply and eventually aligned their growth schedule commensurate with the annual application (Hellmeier et al. 1997). The anticipation of environmental variation for growth either under water or in air by separate morphological programs has already been referred to (Henslow 1908, Schmalhausen 1949). Such programs indicate an ability to anticipate environmental change, even though it may not happen during the lifetime of the individual plant.

There are many perennial plants that live in seasonal environments where the developmental commitments of meristems to vegetative or reproductive structures take place months and years before the organs are elaborated (Geber et al. 1997). Detailed examination of one such plant, the Mayapple, indicates that two decisions are taken at least one year ahead and in some individuals two years. These are the decision on whether the rhizome should branch and the determination of next year's shoot type (vegetative or reproductive). Investigations of the environmental and developmental information that goes to inform these decisions are complex, involving the current resource status, the ability to acquire future resources, and the current resource expenditures on reproduction or growth. The molecular basis of such decision making is not understood, but it enables the Mayapple to exploit its habitat on the forest floor successfully and intelligently.

Plant Learning and Memory

The simplest way to examine whether an organism can learn is to impose circumstances that it cannot have experienced during evolution and observe how it responds with time. It can be anticipated that, initially, such circumstances might impair growth, followed by a recovery and perhaps increased growth as the plant learns to deal with the new situation. There are many such experiments in the literature in which plants respond to treatment with a range of artificial chemicals (herbicides, organic solvents, respiratory inhibitors SH group reagents, etc.) and can respond either by increasing root production and increasing shoot growth or by breakage of seed and bud dormancy (Appleby 1998; Townsend 1897; Trewavas 1992). In terms of growth, the initial effect of an herbicide such as phosfon D may be a reduction in growth, but this is often followed later by an overcompensation, with final growth rates very much higher than controls (Calabrese and Baldwin 2001).

Furthermore, an evident alternative to demonstrating learning is the imposition of stress, such as high cadmium or salt (osmotic stress), low or high temperature, mechanical stress, or very low mineral levels (Amzallag et al. 1990; Baker et al. 1985; Brown and Martin 1981; Henslow 1895; Ingestad and Lund 1979; Laroche et al. 1992; Zhong and Dvorak 1995). If the immediate stress is severe, death or mechanical breakage is the usual result. But if the stress is increased gradually with time, stress conditions that would normally have killed can now be imposed and still permit continued development. Nitrogen stress is particularly instructive since a clear period of radical change and learning in young seedlings is undergone before growth at a lower nitrogen level is then achieved (Ingestad and Lund 1979). Clearly, the plant has learned how to deal with the stressful state and adjust the metabolism and internal structure to cope. Is this any different to teaching *Drosophila* simple avoidance behavior by means of electric shocks?

Many examples of plant memory exist (Desbiez et al. 1984, 1991; Jaffe and Shotwell 1980; Lam and Leopold 1961; Marx 2004; Verdus et al. 1997). A number of plant responses requires two signals for completion. Since each signal will have some unique signal transduction pathways of information flow but clearly later integrate, separating the imposition of the signals in time enables estimates to be made of the length of memory of each signal. Such research referenced above indicates that individual signals can be remembered for minutes, hours, days, or even years. Other examples of memory are to be found in Trewavas (2003).

Organizational Analogies between Trees and Social Insects

There a number of organizational similarities between plants (trees, in particular) and social insect colonies.

- Both trees and colonies contain large numbers of replaceable foragers: in the hive, for example, individual bees (Seeley 1995), in the tree leaf or branch root, meristems.
- In both cases, reproductive and other functions are differentiated from the same uniform genetic line.
- The hive colony is aggressive to invading outsiders, and entry points are guarded. Trees use allelopathy to damage local competitive species and possess induced defense reactions, such as natural pesticides, to kill herbivores or invading fungi (Karban and Baldwin 1997). These defense reactions can be complex, involving chaotic pesticide production in different leaves so that the herbivore is uncertain whether the next leaf is edible or whether consumption kills (Karban and Baldwin 1997).
- A good source of food attracts more insect workers through positive feedback mechanisms and communication. Tree branches and leaves grow to exploit light patches, and roots proliferate in mineral-rich zones involving positive feedback mechanisms and communication (Aphalo and Ballare 1995).
- Just as entry guards to hives and other foraging individuals in hives will altruistically sacrifice themselves to maintain the whole colony and, in particular, the queen, trees will altruistically abscise their foraging organs when parasitized by disease or damaged by herbivores. The abscission zone, a layer of a few cells at the base of the petiole and able to secrete cell wall weakening hydrolytic enzymes, will do so when signals are received from elsewhere in the plant and the leaf blade to commence abscission. The aim is maintenance of the whole individual for later reproduction. Again, leaves and roots can altruistically abscise if resources of minerals and water are short to ensure the future integrity of the individual plant.
- Hive and tree behaviors are dependent on complex communication, assessment of external status, and behavioral (plasticity) change. If one is regarded as intelligent, so must the other.

Summary on Plant Intelligence

Plants exhibit all the characteristics of intelligent behavior described previously. They perceive their environment in considerable detail, make meaningful assessments of that information, and institute adaptive phenotypic responses designed to improve competitive ability and resource acquisition. In addition, future predictions of resource availability can be made and the necessary action taken to reduce or mitigate the problems that might then occur.

Plant intelligence is clearly decentralized and involves the whole organism in the assessment. Like other forms of intelligent activity described above, growing plants are a network of cells, tissues, and organs in which information flow through the network can be altered by environmental change; learning and memory also provide for intelligent behavior. Intelligence is an emergent property that results from complex interactions between the tissues and cells of the individual plant (Trewavas 2004). Peak et al. (2004) have shown clearly how stomata on the leaf surface, physically separated from each other by other epidermal cells, nevertheless integrate into local patches of more uniform behavior; by further integration of the patches, gas exchange is optimized throughout the whole leaf. Although physically separated from each other, stomata form a sparsely connected network over the whole leaf surface, but the final tissue (leaf) behavior is an emergent property dependent in particular on the interactions that occur between the individual cells and patches. Patchiness in behavior of groups of plants cells has been observed several times (references in Trewavas 2003), and the same conclusion operates: intelligence emerges as a result of complex interactions between the constituents of the network.

Communication underpins intelligence in plants. We now know that enormous numbers of specific chemical signals are circulated through the plant body, including proteins, peptides, nucleic acids, oligonucleotides, oligosaccharides, and a plethora of other small molecules, hormones, nutrients, gases, etc. (Trewavas 2003). Mechanical signaling is also crucial because plants are interlinked mechanical structures and under continuous tension from cellular turgor pressures of varying strengths. All such signals have been demonstrated to have specific influences on metabolism, growth, and development.

Perhaps the most revealing example of communication involves grafting. It is known that, for many fruit trees, different root stocks can change shoot phenotypic traits. The current list of alterable shoot traits includes general branching habit, height, fruit bud formation, yield, winter hardiness, disease resistance, and leaf color. Specific homeobox proteins are transferred from root to shoot accounting for some of these observations (Kim et al. 2001). Complex signalling between the root and the shoot integrates the whole organism and enables critical and necessary alterations in information flow as environmental signals change.

Evolution and the Convergence of Intelligence

The discussion above indicates that intelligent behavior can be found throughout biology from bacteria to plants to animals. Clearly, intelligence is an excellent example of convergence, the phenomenon elegantly described by Conway Morris (2003). Animals evolved nerve cells to speed up connections between sensory and response systems and to compete successfully with other individuals in outdoing either predators or prey. The evolution of nerve cells with numerous dendrite connections improved the accuracy of assessment and memory, and with more complex brains, emergent properties (such as cognition) developed. Analogously, quorum sensing, that is, increased communication between bacterial cells, generates emergent properties of luminescence or virulence when sufficient numbers of cells are involved.

However, other organisms have developed intelligence in a very different way from animals. Here, intelligence does not localize in a defined place like a brain but is a property of the whole system. Animals learn by exchanging dendritic connections between different cells, constructing new neural pathways and changing information flow. Analogously, bacteria learn by exchanging genes from other bacteria, altering information flow. Cells learn by changing directions of information flow through signal transduction pathways. Plants learn by changing information flow via chemical communication much as social insects do. Individual cells are capable of computation, but multicellular organisms, composed as they are of communicating cells, should be capable of much greater degrees of assessment and memory.

Underpinning all the forms of intelligence described in this chapter is a network whose connection strength can be altered, enabling control of information flow and memory to be constructed. The components of the network can be individual organisms (species, swarm, bacteria), cells (brain, plant, bacteria, fungi, protozoa), or molecules. Plants, like bacteria, can use exchange of genes through sexual reproduction with large populations of genetically variable individuals. But plants have gone one better than bacteria, using substantive phenotypic (morphological) plasticity as well.

Why, then, is intelligence so widely found in biology? The critical issue that generates intelligent organisms is surely natural selection in a highly variable environment. Even though each environment will probably be unique to each species, attempts to deal with environmental variability generate similar and, thus, convergent solutions, resulting in intelligent behavior. If the environment is (in evolutionary terms) stable, then intelligent solutions are not necessary, and autonomic responses can satisfy the requirements for food and mating.

Intelligent Behavior Is Used to Help Stabilize the Conclusion of the Life Cycle

Homeostasis is a term originally used to describe the ability of animals to maintain a relatively constant blood pH, temperature, ionic strength, fluid amount, oxygen, and various blood constituents, like calcium and fat content (Cannon 1932). Homeostatic mechanisms operate by negative feedback comparing present status to a predetermined goal or setting. After perturbation, the approach to this predetermined value will be a series of damped oscillations around the predetermined goal. In this respect, homeostasis has many similarities to learning behavior in animals. Usually, there are a prespecified goal and an error-correcting mechanism in learning that assesses how remote is current behavior from the desired objective (Trewavas 2003).

Cells also have mechanisms to stabilize the intracellular environment. There are feedback (homeostatic) mechanisms to control the flux through numerous metabolite pathways, as well as regulations of ionic strength, ion flux, intracellular pH, and the levels of numerous proteins. Perturbations of any of these cellular constituents as a result of environmental change can then initially be used as signals eliciting the operation of control mechanisms (usually via protein phosphorylation) to bring the cell back into the original steady state. If the perturbations are too large, the fluxes through the metabolite and protein network will settle into a new stabilized configuration, and new proteins will be synthesized, eliciting phenotypic changes in cell behavior and perhaps a new developmental pathway. Analogously, here are the different mechanisms that underpin short- and long-term memory in brains. Short-term memory involves temporary changes in ion fluxes usually via protein phosphorylation. Long-term memory involves the formation of new dendrites and, thus, a new network structure (Trewavas 2003).

Homeostasis must have been one of the earliest requirements of the primordial cell (Trewavas 1988). Without intracellular, homeostatic regulation, protein denaturation would have been frequent and destructive. A constant cell environment is maintained because enzymes with optimal activities can then evolve, enabling a more efficient metabolism and success against competitive rivals. By analogy, intelligent behavior is surely an attempt to ensure homeostasis of the life cycle! Achievement, accomplishment, of a more reliable life cycle than competitors will result in greater numbers of progeny and, thus, greater fitness. Those best able to master their environment intelligently are those most likely to succeed and pass on genes into the next generation.

Competition between species may be the reason for overall species intelligence as described by Schull (1990); species with the greater intelligent capability will out-perform different species attempting to occupy the same ecological niche and foraging for the same food. All organisms face the same basic requirement to find food, whether as autotrophs or heterotrophs. Much intelligent behavior is directed towards successful foraging when the food supply cannot be guaranteed and is extremely variable. If the food supply were abundant and completely reliable, intelligent behavior would no longer be required to forage successfully. Attempts to mitigate the variable environment and compete at the same time lead to the convergent evolution of intelligent behavior, regardless of the specifics of the particular environment concerned.

I have emphasized the need for plant foraging in this chapter because attempts to optimize foraging and to stabilize a successful life cycle are crucial to understanding plant behavior, defined here as phenotypic plasticity. Food for plants is just as variable in the appropriate environment as it is for animals. Whereas animals can move to solve environmental difficulties, the sessile plant must rely on changes in structure. Cannon (1932) indicated how the pressures of thirst and hunger in higher animals are used to help maintain internal homeostasis. The same homeostatic imperative to find light and/or water, when either is short, is found in many plants. For mild lack of water, minerals, or light, the plant response involves changes in flux rates of uptake of nutrients via temporary changes (by phosphorylation) in proteins and metabolism. More severe shortages necessitate changes in phenotype. Note the similarity between short- and long-term memory in the brain indicated above.

Behavior, Intelligence, and Genetic Assimilation

A shift into a new niche or adaptive zone requires almost without exception a change in behavior. It is now quite evident every habit and behavior has some structural basis but that the evolutionary change that result from adaptive shifts are often initiated by changes in behavior to be followed secondarily by a change in structure. It is *very often* the new habit which sets up the selection pressure that shifts the mean curve of structural variation.

Mayr 1960, 371

It is not the organs . . . of an animal's body that have given rise to its special habits and faculties: but it is on the contrary, its habits, mode of life and environment that have in the course of time controlled the shape of its body, the number and state of its organs and lastly the faculties which it possesses.

Lamarck 1809, 114

There are currently at least two kinds of evolutionary models relevant to this discussion (Jablonka and Lamb 1995). The first, the neo-Darwinian view, sees overproduction, random genetic variation, and differential survival as the basis of evolution (e.g., Gould 2002). The second, best described as "the survival of the adaptable" (Waddington 1957) or genetic assimilation (among other terms), is often described as neo-Lamarckian (Jablonka and Lamb 1995). This mechanism places behavioral changes as the first response to environmental shifts, as indicated in the two quotations from Mayr and

Lamarck above. How far, then, does phenotypic plasticity in plants, indicated here to be plant behavior, fit with this perspective?

The mechanism of genetic assimilation differs substantially from the neo-Darwinian view. Environmental shifts, it is suggested (Mayr 1960), institute adaptive changes among the population. The ability to adapt will be variable among the population as indicated by norms of reaction (Schlichting and Pigliucci 1998). Those that adapt most efficiently and are, thus, best able to master the current changes in the environment will experience preferential survival. However, these individuals will represent a discrete subset of the genetic variation in the population (Waddington 1957). If the novel situation persists, continued interbreeding among the successful individuals (which now become a deme) should promote the natural selection of genes and gene combinations that allow the strategy to develop with greater rapidity, higher probability, or lower cost (Bateson 1963). Eventually, individuals in which the adaptive change is permanently expressed (and thus assimilated) arise because the continued environmental shift no longer provides benefit for the characteristics to be adaptable. Thus, for genetic assimilation, natural selection ratifies an adaptation that has already been tested and developed through nongenetic means (Jablonka and Lamb 1995). No view of evolution can exclude either the neo-Darwinian or neo-Lamarckian mechanisms, but the conditions under which either becomes the dominant mechanism may be very different.

Genetic assimilation provides a direction and speed to evolution: it emphasizes the directing role of the environment in evolution and is, in contrast to the reductionist theme of the neo-Darwinian synthesis, based only on upward causation from genes. Genetic assimilation is a clear example of downward causation, in which both the genetic and epigenetic components of the organism and indeed probably the whole life cycle are subject to competitive selection (Corning 2003; Jablonka and Lamb 1995; Schlichting and Pigliucci 1998). Whatever genes the successful organism possesses go along for the ride, as it were (Corning 2003). Sperry's description of downward causation used that of a spoked wheel (Corning 2003). Although the wheel runs downhill on its rim, all the components of the wheel (spokes and axle) have to go down with it.

Genetic Assimilation in Plant Evolution

Angiosperm plants are believed to have originated about 125 million years ago, and there are now an estimated 250,000 distinct species, very many more than the gymnosperms and ferns that preceded them. In my view, this rapidity of speciation could have resulted from the dominance of genetic assimilation mechanisms. The success of the primordial angiosperms arose from a much greater potential for phenotypic and physiological plasticity. Thus, for example, in comparison to gymnosperms, angiosperm branching patterns, leaf shape, and size are notably plastic; control of water loss is much more refined; and pollen tube growth takes a few days instead of years. But plasticity in structure and in morphology, in turn, generates a more complex mosaic of available resources to other individuals. Each successive new species, then, arose by refining its discrimination of the resource mosaic and producing both structure and physiology to optimize resource acquisition competitively under these circumstances. However, as a result, the resource mosaic becomes yet more complex. Rapid speciation has, thus, arisen from a positive feedback of individual plants upon each other.

Much animal speciation seems to depend on either acquiring a new organism for food or a new feeding mechanism (e.g., cichlid fishes) to acquire more competitively the same food as others. By this means, a new ecological niche is generated. Analogously, plant speciation depends on an increasing competitive refinement in resource sensing in an increasingly complex resource mosaic and generating the necessary refinements in phenotypic change to exploit. Thus, the number of discrete but different environmental circumstances for plant exploitation becomes as large as the number of different animals to act as food for animal predators. Neighbor identity in plants is a major source of specific phenotypic variability (Callaway et al. 2003; Huber-Sannwald et al. 1997; Turkington and Klein 1991; Turkington et al. 1991). Not only does the specific neighbor institute specific phenotypic changes in the individual, but the effects of a particular neighbor can be remembered for considerable periods of time after the neighbor has disappeared. Those best able to master this environmental complexity and reproduce will see an increasing progeny survival in competition with others.

Henslow (1895, 1908) drew attention to the many examples of clear adaptive character in plants sufficient to explain many cases of evolution. Thus, for example, *Ampelopsis hederacea* (Virginia creeper) only forms tendril pads upon contact, a clear response to a mechanical stimulus and resulting from phenotypic plasticity. *Ampelopsis veitchii* partially forms tendril pads before mechanical contact; some other *Ampelopsis* species do not form tendril pads at all but climb only by coiling tendrils. The obvious conclusion is that *A. veitchii* evolved from *A. hederacea*; but because the requirement to climb walls was so common, the adaptive character has now become fixed in forming the new species by genetic assimilation. There is a surprising evolutionary parallel in animals described by Waddington (1957). In the ostrich, callosities form exactly where the sitting bird touches the ground. Although callosities normally form upon continuous skin stimulation, these callosities are seen in the fetus in the egg. Thicker skin found on the soles of the human feet are an adaptive feature but are also found in the human fetus (Waddington 1957).

Genetic assimilation was also described by Schmalhausen (1949), using instead the term *stabilizing selection*. Schmalhausen illustrates stabilizing selection with examples of water plants that anticipate variable environments by the potential production of two completely different phenotypes. Henslow (1908) also describes closely related water plant species to these dual phenotype plants in which only one or other of the two phenotypes is now expressed, regardless of growth in water or land, a clear suggestion of genetic assimilation mechanisms again. Earlier versions of genetic assimilation, although more limited in explanation, were produced by Baldwin (1896) and, as the quotation indicates above, by Lamarck in 1809. Both these authors were concerned to indicate the importance of behavioral (habit) changes as molding the subsequent phenotype, in Baldwin's case, neural mechanisms.

There are many examples in plants that can be explained by genetic assimilation that are also examples of the concept of convergent evolution (Conway Morris 2003). There are convergent examples in the form and habit of desert plants such as the Cactaceae and Euphorbiacae (Conway Morris 2003). Although both groups originated on different continents, they share many common structural characteristics of large photosynthetic stem and spines (as well as similar internal structure) but can be distinguished because Euphorbias produce vestigial leaves. Schmalhausen (1949) points out that temperate plants, such as bean, will undergo the behavioral changes of stem thickening and enhanced stem greening as an adaptive response if all the leaves are removed. Other plants will form spines or leaves as an adaptive behavioral and phenotypically plastic feature, dependent on water supply (Henslow 1908; Sinnott 1960). Such changes mimic some of the primary structural characteristics of the Euphorbiaceae/Cactaceae and suggest that genetic assimilation may have underpinned their evolution.

Other convergent examples are to be found in plants from many different genera that, nevertheless, exhibit similar xeromorphic characteristics. Xeromorphy results from reductions in water supply or from imposition of salt stress from sea spray. These phenotypic changes are illustrated by the structure of newly developed leaves. These leaves frequently become succulent and have a reduced internal transpirational surface and vasculature, reduced stomatal density, thicker cuticles and increased hairiness, and conversion of leaves to spines (Henslow 1908; Hsaio et al. 1976; Sinnott 1960; Stocker 1960). There are species that live permanently near the sea (e.g., sea holly) or in deserts in which these characteristics are no longer adaptive. Arctic alpines of many different kinds share a dwarf stature and increased hairiness, a response to increased wind stimulation at altitude (Waddington 1957). At lower altitudes, these plants can grow taller in more optimal growth conditions but never achieve the height of equivalent lowland species. Furthermore, poor circumstances for growth will result in dwarfing of lowland plants. The dwarf characteristic is clearly partly inherited and partly adaptive. Climbing plants are characterized by strong reductions in the main strengthening tissues and polymers such as lignin. Henslow (1908) pointed out that circumnutation can be used to enable some ground-covering plants to climb around poles, but not all circumnutating plants will do so. Plants that can grow under water or on land increase their amount of aerenchyma and reduce vascular tissue to allow passage of oxygen from the surface leaves to the roots when under water. Buoyancy reduces the need for structural tissues, and being surrounded by water reduces the need for water-conducting vascular tissues. The figwort possesses many of these characteristics, although it does not grow under water (Henslow 1908). Henslow (1895) suggests that the structure of certain flowers was initially an adaptive character determined by the weight of a pollinating insect's alighting on them. The critical event in these examples above seems to be a prior behavioral change in the phenotype; only later will such changes become fixed in other species.

Each of these phenotypic traits or characters described above results from an adaptation to environmental stress, such as reductions in available water, light, low oxygen, continual wind stimulation, and/or low temperature at altitude. Convergence in evolution of water plants, xeromorphic plants, or arctic alpines should, then, occur when the resource requirement is overwhelming, such that considerable stress is imposed upon the organism that is attempting to move into the environment. The evolution of such similar traits in different species may reflect the more constrained capabilities of adaptation. But such similarity enhances the claims of genetic assimilation to be a major evolutionary mechanism in stressful circumstances. The very refined and complex forms of innate behavior found in reproductive rituals in animals and birds must surely originally have been learned behavior that has now been genetically assimilated. Competition to acquire mates provided the necessary stress.

Intelligent Behavior Is a Critical Feature in Plant Evolution

Genetic assimilation is initiated by changes in behavior, and, in the plant examples above, behavior is expressed as phenotypic plasticity, which I have indicated is intelligent behavior. Intelligent behavior is, thus, a critical trait that is selected and developed particularly in angiosperms because of an increasing complexity in the resource mosaic. *The evolution of intelligent behavior found in all forms of life, thus, becomes a central theme in the evolution of life itself.*

But the mechanism whereby intelligence is expressed has changed. The decentralized bacterial intelligence involves a population of single and sometimes cohering cells dependent on genetic variation and communication via rapid exchange of genes between species. In contrast, in primates, intelligence is located in one or two tissues and is a property of the individual organism, involving rapid chemical or electrical communication. Perhaps the comparison is unfair in the sense that a primate is a conglomerate of billions of closely addressed cells and in which speed of lifecycle and of resource acquisition necessitates rapid communication between cells. But plants do come somewhere in between these two extremes. Their intelligence, such as it is, involves the whole plant and is decentralized, is dependent on chemical communication but still involves the movement of transcripts between cells and organs as well as other molecules. Furthermore, plants can sense self and nonself and actively change phenotype when competition from alien individuals or species is experienced. Whatever the difference, the convergence of intelligent mechanism is a striking property, little commented on in evolutionary circles, but it now demands further investigation. Those best able to master the environment will become the most fit, and mastery is ultimately dependent on intelligence.

References

Aarsen, L. W. 1995. Hypotheses for the evolution of apical dominance in plants: Implications for the interpretation of overcompensation. *Oikos* 74: 149–56.

Ackerley, D. D., and F. A. Bazzaz. 1995. Seedling crown orientation and interception of diffuse radiation in tropical forest gaps. *Ecology* 76: 1134–46.

Addicott, F. T. 1982. *Abscission.* Berkeley: University of California Press.

Agrawal, A. A., C. Laforsch, and R. Tollrian. 1999. Transgenerational induction of defences in animals and plants. *Nature* 401: 60–63.

Amzallag, G. N., H. R. Lerner, and A. Poljakoff-Mayber. 1990. Induction of increased salt tolerance in *Sorghum bicolor* by sodium chloride treatment. *Journal of Experimental Botany* 41: 29–34.

Aphalo, P. J., and C. L. Ballare. 1995. On the importance of information-acquiring systems in plant-plant interactions. *Functional Ecology* 9: 5–14.

Appleby, A. P. 1998. The practical implications of hormetic effects of herbicides on plants. *Human and Experimental Toxicology* 17: 270–71.

Arkin, A., and J. Ross. 1994. Computational functions in biochemical reaction networks. *Biophysical Journal* 67: 560–78.

Attenborough, D. 1995. *The Private Life of Plants.* BBC Natural History Unit, British Broadcasting Corporation, London. TV production in association with Turner Broadcasting Systems Inc., London.

Baillaud, L. 1962. Mouvements autonomes des tiges, vrilles et autre organs. In *Physiology of movements.* Vol. XVII, part 2 of *Encyclopedia of plant physiology,* ed. W. Ruhland, 562–635. Berlin: Springer-Verlag.

Baker, A. J. M., C. J. Grant, M. H. Martin, S. C. Shaw, and J. Whitebrook. 1985. Induction and loss of cadmium tolerance in *Holcus lanatus* and other grasses. *New Phytologist* 102: 575–87.

Baldwin, J. M. 1896. A new factor in evolution. *American Naturalist* 30: 441–51.

Ballare, C. L. 1994. Light gaps: Sensing the light opportunities in highly dynamic canopy environments. In *Exploitation of environmental heterogeneity by plants*, ed. M. M. Caldwell and R. W. Pearcy, 73–111. New York: Academic Press.

———. 1999. Keeping up with the neighbours: Phytochrome sensing and other signalling mechanisms. *Trends in Plant Sciences* 4: 97–102.

Bassler, B. L. 2002. Small talk: Cell-to-cell communication in bacteria. *Cell* 109: 421–24.

Bateson, G. 1963. The role of somatic change in evolution. *Evolution* 17: 529–39.

Bazzaz, F. A. 1991. Habitat selection in plants. *American Naturalist* 137: S116–S130.

———. 1996. *Plants in changing environments.* Cambridge: Cambridge University Press.

Bell, A. D. 1984. Dynamic morphology: A contribution to plant population ecology. In *Perspectives on plant population ecology*, ed. R. Dirzo and J. Sarukhan, 48–66. Sunderland, MA: Sinauer Associates, Inc.

Bell, G., and M. J. Lechowicz. 1994. Spatial heterogeneity at small scales and how plants respond to it. In *Exploitation of environmental heterogeneity by plants*, ed. M. M. Caldwell and R. W. Pearcy, 391–411. New York: Academic Press.

Bloom, A. J., F. S. Chapin, and H. A. Mooney. 1985. Resource limitation in plants—an economic analogy. *Annual Review of Ecology and Systematics* 16: 363–92.

Bonabeau, E., M. Dorigo, and G. Theraulax. 2000. Inspiration for optimisation from social insect behavior. *Nature* 406: 39–42.

Bonabeau, E., and C. Meyer. 2001. Swarm intelligence. *Harvard Business Review* 79(5): 107–14.

Bonabeau, E., and G. Theraulaz. 2000. Swarm smarts. *Scientific American* 282: 72.

Bradshaw, A. D. 1965. Evolutionary significance of phenotypic plasticity. *Advances in Genetics* 13: 115–55.

Bray, D. 1995. Protein molecules as computational elements in living cells. *Nature* 376: 307–12.

Brown, H., and M. H. Martin. 1981. Pre-treatment effects of cadmium on the root growth of *Holcus lanatus*. *New Phytologist* 89: 621–29.

Cairns, J., J. Overbaugh, and S. Miller. 1988. The origin of mutants. *Nature* 335: 142–45.

Calabrese, E. J., and L. A. Baldwin. 2001. Hormesis: U-shaped dose responses and their centrality in toxicology. *Trends in Pharmacological Sciences* 22: 285–91.

Callaway, R. M., S. C. Pennings, and C. L. Richards. 2003. Phenotypic plasticity and interactions among plants. *Ecology* 84: 1115–28.

Cannon, W. B. 1932. *The wisdom of the body.* New York: W. W. Norton and Co.

Charnov, E. L. 1976. Optimal foraging, the marginal value theorem. *Theoretical Population Biology* 9: 129–36.

Conway Morris, S. 2003. *Life's solution. Inevitable humans in a lonely universe.* Cambridge: Cambridge University Press.

Corning, P. 2003. *Nature's magic-synergy in evolution and the fate of humankind.* Cambridge: Cambridge University Press.

Coutinho, A. 2002. How evolution of development tinkered the emergence of complex behaviours in the immune system. http://www.c3.lanl.gov/-rocha/embrob/coutinho.html

Darwin, C. 1882. *The power of movement in plants.* London: John Murray.

———. 1891. *The movements and habits of climbing plants.* London: John Murray.

De Castro, L. N., and J. I. Timmis. 2002. *Artificial immune systems: A new computational intelligence approach.* London: Springer-Verlag.

De Kroon, H., and M. J. Hutchings. 1995. Morphological plasticity in clonal plants: The foraging concept reconsidered. *Journal of Ecology* 83: 143–52.

Desbiez, M. O., Y. Kergosein, P. Champagnant, and M. Thellier. 1984. Memorisation and delayed expression of regulatory message in plants. *Planta* 160: 392–99.

Desbiez, M. O., M. Tort, and M. Thellier. 1991. Control of a symmetry breaking process in the course of morphogenesis of plantlets of *Bidens pilosa*. *Planta* 184: 397–402.

Drew, M. C., and L. R. Saker. 1975. Nutrient supply and the growth of the seminal root system in barley. *Journal of Experimental Botany* 26: 79–90.

Durrant, A. 1962. The environmental induction of heritable change in *Linum*. *Heredity* 17: 27–61.

———. 1981. Unstable genotypes. *Philosophical Transactions of the Royal Society of London*. Series B 292: 467–74.

Edelman, G. M. 1993. Neural Darwinism: Selection and re-entrant signalling in higher brain function. *Neuron* 10: 115–25.

Evans, J. P., and M. L. Cain. 1995. A spatially explicit test of foraging behavior in a clonal plant. *Ecology* 76: 1147–55.

Falik, O., P. Reides, M. Gersani, and A. Novoplansky. 2003. Self, non-self discrimination in roots. *Journal of Ecology* 91: 525–31.

Franco, M. 1986. The influence of neighbours on the growth of modular organisms with an example from trees. *Philosophical Transactions of the Royal Society of London*. Series B 313: 209–25.

Franks, N. R., A. Dornhaus, J. P. Fitzsimmons, and M. Stevens. 2003. Speed versus accuracy in collective decision-making. *Proceedings of the Royal Society of London*. Series B 270: 2457–63.

Gavin, A. C., M. Bosche, R. Krause et al. 2002. Functional organisation of the yeast proteome by systematic analysis of protein complexes. *Nature* 415: 541–47.

Geber, M. A., M. A. Watson, and H. De Kroon. 1997. Organ preformation, development and resource allocation in perennials. In *Plant resource allocation*, ed. F. A. Bazzaz and J. Grace, 113–43. London: Academic Press.

Gersani, M., Z. Abramsky, and O. Falik. 1998. Density-dependent habitat selection in plants. *Evolutionary Ecology* 12: 223–34.

Gersani, M., J. S. Brown, E. E. O'Brien, G. M. Maina, and Z. Abramsky Z. 2001. Tragedy of the commons as a result of root competition. *Ecology* 89: 660–69.

Gersani, M., and T. Sachs. 1992. Developmental correlations between roots in heterogenous environments. *Plant Cell and Environment* 15: 463–99.

Gilroy, S., and A. J. Trewavas. 2001. Signal processing and transduction in plant cells: The end of the beginning? *Nature Molecular Cell Biology Reviews* 2: 307–14.

Givnish, T. J. 1982. On the adaptive significance of leaf height in forest herbs. *American Naturalist* 120: 353–81.

Gleeson, S. K., and J. E. Fry. 1997. Root proliferation and marginal patch value. *Oikos* 79: 387–93.

Goldberg, D. E., and A. M. Barton. 1992. Patterns and consequences of interspecific competition within natural communities: A review of field experiments with plants. *American Naturalist* 139: 771–801.

Gould, S. J. 2002. *The structure of evolutionary theory*. Harvard, MA: Harvard University Press.

Granato, T. C., and C. D. Raper. 1989. Proliferation of maize roots in response to localised supply of nitrate. *Journal of Experimental Botany* 40: 263–75.

Greengard, P. 2001. The neurobiology of slow synaptic transmission. *Science* 294: 1024–30.

Grime, J. P. 1994. The role of plasticity in exploiting environmental heterogeneity. In *Exploitation of environmental heterogeneity by plants*, ed. M. M. Caldwell and R.W. Pearcy, 1–19. New York: Academic Press.

Grime, J. P., J. C. Crick, and J. E. Rincon. 1986. The ecological significance of plasticity. In *Symposium of the Society for Experimental Biology and Medicine* XL. *Plasticity in plants*, ed. D. H. Jennings and A. J. Trewavas, 5–29. London: Cambridge University Press.

Groseburg, R. K., and M. W. Hart. 2000. Mate selection and the evolution of highly polymorphic self/non self recognition genes. *Science* 289: 2111–14.

Gruntman, M., and A. Novoplansky. 2004. Physiologically mediated self/non self discrimination mechanism. *Proceedings of the National Academy of Sciences* USA 101: 3863–67.

Hall, B. G. 1992. Selection-induced mutations in yeast. *Proceedings of the National Academy of Sciences USA* 89: 4300–4303.

Harper, J. L. 1977. *The population biology of plants*. London: Academic Press.

Hart, M. W., and R. K. Groseburg. 1999. Kin interactions in a colonial hydrozoan: Population structure on a mobile landscape. *Evolution* 53: 793–805.

Hartnett, D. C., and F. A. Bazzaz. 1983. Physiological integration among intra-clonal ramets in *Solidago canadensis*. *Ecology* 64: 779–88.

Hellmeier, H., M. Erhard, and E. D. Schulze. 1997. Biomass accumulation and water use under arid conditions. In *Plant resource allocation*, ed. F. A. Bazzaz and J. Grace, 93–113. London: Academic Press.

Henrikkson, J. 2001. Differential shading of branches or whole trees: Survival, growth and reproduction. *Oecologia* 126: 482–86.

Henslow, G. 1895. *The origin of plant structures by self adaptation to the environment*. London: Kegan Paul, French, Trubner and C., Ltd. (The adaptation to mechanical stress is described on page 204 but is derived from a description by Pfeffer).

———. 1908. *The heredity of acquired characters in plants*. London: John Murray.

Highkin, H. R. 1958. Temperature induced variability in peas. *American Journal of Botany* 45: 626–31.

Hill, J. 1965. Environmental induction of heritable changes in *Nicotiana rustica*. *Nature* 207: 732–34.

Holzapfel, C., and P. Alpert. 2003. Root co-operation in a clonal plant: Connected strawberries segregate roots. *Oecologia* 134: 72–77.

Honda, H., and J. B. Fisher. 1978. Tree branch angle: Maximising effective leaf area. *Science* 199: 888–89.

Honkanen, T., and E. Hanioja. 1994. Why does a branch suffer more after branch-wide than after tree-wide defoliation? *Oikos* 71: 441–50.

Hsaio, T. C., E. Acevedo, E. Fereres, and D. W. Henderson. 1976. Stress metabolism. *Philosophical Transactions of the Royal Society of London*, Series B 273: 479–500.

Huber-Sannwald, E., D. A. Pyke, and M. M. Caldwell. 1997. Perception of neighbouring plants by rhizomes and roots: Morphological manifestations of a clonal plant. *Canadian Journal of Botany* 75: 2146–57.

Hutchings, M. J. 1997. Resource allocation patterns in clonal herbs and their consequences for growth. In *Plant resource allocation*, ed. F. A. Bazzaz and J. Grace, 161–86. London: Academic Press.

Hutchings, M. J., and H. De Kroon. 1994. Foraging in plants, the role of morphological plasticity in resource acquisition. *Advances in Ecological Research* 25: 159–238.

Ingestad, T., and A. B. Lund. 1979. Nitrogen stress in birch seedlings. *Physiologia Plantarum* 45: 137–48.

Jablonka, E., and M. J. Lamb. 1995. *Epigenetic inheritance and evolution*. Oxford: Oxford University Press.

Jackson, R. B., and M. M. Caldwell. 1989. The timing and degree of root proliferation in fertile soil microsites for three cold desert perennials. *Oecologia* 81: 149–53.

Jaffe, M. J., and M. Shotwell. 1980. Physiological studies on pea tendrils. XI. Storage of tactile sensory information prior to the light activation effect. *Physiologia Plantarum* 50: 78–82.

Jennings, D. H., and A. J. Trewavas. 1986. *Plasticity in plants*. Vol. XL of *Symposium of the Society for Experimental Biology and Medicine*. Cambridge: Cambridge University Press.

Jones, M., and J. L. Harper. 1987. The influence of neighbours on the growth of trees. I. The demography of buds in *Betula pendula*. *Proceedings of the Royal Society of London*, Series B 232: 1–18.

Kandel, E. R. 2001. The molecular biology of memory storage. A dialogue between genes and synapses. *Science* 294: 1030–38.

Karban, R., and I. T. Baldwin. 1997. *Induced responses to herbivory*. Chicago: University of Chicago Press.

Kelly, C. K. 1992. Resource choice in *Cuscuta europea*. *Proceedings of the National Academy of Sciences USA* 89: 12194–97.

Kelly, C. L. 1990. Plant foraging: A marginal value model and coiling response in *Cuscuta subinclusa*. *Ecology* 71: 1916–25.

Kim, M., W. Canio, S. Keller, and N. Sinha. 2001. Developmental changes due to long distance movement of a homeo-box fusion transcript in tomato. *Science* 293: 287–93.

Kovalchuk, I., O. Kovalchuk, V. Kalck, V. Boyko, J. Filkowski, M. Heinlein, and B. Hohn. 2003. Pathogen-induced systemic plant signal triggers DNA rearrangements. *Nature* 423: 760–62.

Kumar, A., and J. L. Bennetzen. 1999. Plant retro-transposons. *Annual Review of Genetics* 33: 479–532.

Kuppers, M. 1994. Canopy gaps: Competitive light interception and economic space filling. In *Exploitation of environmental heterogeneity by plants*, ed. M. M. Caldwell and R. W. Pearcy, 111–44. New York: Academic Press.

Lam, S. L., and A. C. Leopold. 1961. Reversion and re-induction of flowering in *Perilla*. *American Journal of Botany* 48: 306–10.

Lamarck, J. B. 1809. *Zoological philosophy. An exposition with regard to the natural history of animals*. Trans. H. Elliott, 1914. London: MacMillan and Co, Ltd.

Laroche, A., X. M. Geng, and J. Singh. 1992. Differentiation of freezing tolerance and vernalisation responses in Cruciferae exposed to a low temperature. *Plant Cell and Environment* 15: 439–46.

Mahall, B. E., and R. M. Callaway. 1992. Root communication mechanism and intra-community distributions of two Mojave desert shrubs. *Ecology* 73: 2145–51.

Maina, G. G., J. S. Brown, and M. Gersani. 2002. Intra-plant versus inter-plant competition in beans: Avoidance resource matching or tragedy of the commons. *Plant Ecology* 160: 235–47.

Marx, J. 2004. Remembrance of winter past. *Science* 303: 1607.

Maslov, S., and K. Sneppen. 2002. Specificity and topology of protein networks. *Science* 296: 910–13.

Mayr, E. 1960. The emergence of evolutionary novelties. In *Evolution after Darwin*, ed. S.Tax, 1:349–80. Chicago: University of Chicago Press.

McClintock, B. 1984. The significance of responses of the genome to challenge. *Science* 226:792–801.

McConnaughay, K. D. M., and F. A. Bazzaz. 1991. Is physical space a soil resource? *Ecology* 72: 94–103.

———. 1992. The occupation and fragmentation of space: Consequences of neighbouring shoots. *Functional Ecology* 6: 711–18.

McGhee, G. R., and F. K. McKinney. 2000. A theoretical morphologic analysis of convergently evolved erect helical colony form in the Bryozoa. *Paleobiology* 26: 556–77.

McKinney, F. K., and G. R. McGhee. 2003. Evolution of erect helical colony form in the Bryozoa: Phylogenetic, functional and ecological factors. *Biological Journal of the Linnean Society* 80: 235–60.

Miller, M. B., and B. L. Bassler. 2001. Quorum sensing in bacteria. *Annual Review of Microbiology* 55: 165–99.

Muth, C. C., and F. A. Bazzaz. 2002a. Tree seedling canopy responses to conflicting photosensory cues. *Oecologia* 132: 197–204.

———. 2002b. Tree canopy displacement at forest gap edges. *Canadian Journal of Forestry Research* 32: 247–54.

———. 2003. Tree canopy displacement and neighbourhood interactions. *Canadian Journal of Forestry Research* 33: 1323–30.

Nakagaki, T., H. Yamada, and A. Toth. 2000. Maze solving by an amoeboid organism. *Nature* 407: 470.

Novick, A., and M. Weiner M. 1957. Enzyme induction as an all-or-none phenomenon. *Proceedings of the National Academy of Sciences USA* 43: 553–66.

Novoplansky, A. 1996. Hierarchy establishment among potentially similar buds. *Plant Cell and Environment* 19: 781–86.

———. 2003. Ecological implications of the determination of branch hierarchies. *New Phytologist* 160: 111–18.

Novoplansky, A., D. Cohen, and T. Sachs. 1989. Ecological implications of correlative inhibition between plant shoots. *Physiologia Plantarum* 77: 136–40.

———. 1990. How *Portulaca* seedlings avoid their neighbours. *Oecologia* 82: 490–93.

Okamoto, M., T. Sakai, and K. Hayashi. 1987. Switching mechanism of a cyclic enzyme system: Role as a chemical diode. *Biosystems* 21: 1–11.

Park, S., P. M. Wolanin, E. A. Yuzbashyan, P. Silberzan, J. B. Stock, and R. H. Austin. 2003a. Motion to form a quorum. *Science* 301: 188.

Park, S., P. M. Wolanin, E. A. Yuzbashyan, H. Lin, N. C. Darnton, J. B. Stock, P. Silberzan, and R. H. Austin. 2003b. Influence of topology on bacterial social interaction. *Proceedings of the National Academy of Sciences USA* 100: 13910–15.

Peak, D., J. D. West, S. M. Messenger, and K. A. Mott. 2004. Evidence for complex collective dynamics and emergent-distributed computation in plants. *Proceedings of the National Academy of Sciences USA* 101: 981–22.

Pearcy, R. W., R. L. Chardin, L. J. Gross, and K. A. Mott. 1994. Photosynthetic utilisation of sunflecks: A temporally patchy resource on a time scale of seconds to minutes. In *Exploitation of environmental heterogeneity by plants*, ed. M. M. Caldwell and R. W. Pearcy, 175–209. New York: Academic Press.

Perkel, J. M. 2004. Validating the interactome. *The Scientist* 18: 19–22.

Pink, D., and I. Puddephat. 1999. Deployment of disease resistance genes by plant transformation-a mix and match approach. *Trends in Plant Science* 4: 71–75.

Ravasz, E., A. L. Somera, D. A. Mongru, Z. N. Oltvai, and A. L. Barabasi. 2002. Hierarchical organisation of modularity in metabolic networks. *Science* 297: 1551–55.

Ries, G., W. Heller, H. Puchta, H. Sandermann, H. K. Seitlitz, and B. Hohn. 2000. Elevated UV-B radiation reduces genome stability in plants. *Nature* 406: 98–101.

Robertson, G. P., and K. L. Gross. 1994. Assessing the heterogeneity of below ground resources: Quantifying pattern and scale. In *Exploitation of environmental heterogeneity by plants*, ed. M. M. Caldwell and R. W. Pearcy, 237–53. New York: Academic Press.

Sachs, T., A. Novoplansky, and D. Cohen. 1993. Plants as competing populations of redundant organs. *Plants, Cell and Environment* 16: 765–70.

Salzman, A. G. 1985. Habitat selection in a clonal plant. *Science* 228: 603–4.

Salzman, A. G., and M. Parker. 1985. Neighbours ameliorate local salinity stress for a rhizomatous plant in a heterogeneous environment. *Oecologia* 65: 273–77.

Schenk, H. J., R. M. Callaway, and B. E. Mahall. 1999. Spatial root segregation: Are plants territorial? *Advances in Ecological Research* 28: 145–80.

Schieving, F., and H. Poorter . 1999. Carbon gain in a multispecies canopy: The role of specific leaf area and photosynthetic nitrogen use efficiency in the tragedy of the commons. *New Phytologist* 143: 201–11.

Schlichting, C. D., and M. Pigliucci. 1998. *Phenotypic evolution: A reaction norm perspective*. Sunderland, MA: Sinauer Associates Inc.

Schmalhausen, I. I. 1949. *Factors of evolution*. Philadelphia: Blakiston.

Schull, J. 1990. Are species intelligent? *Behavioral and Brain Sciences* 13: 63–108.

Seeley, T. D. 1995. *The wisdom of the hive: The social physiology of honey bee colonies.* Cambridge, MA: Harvard University Press.

Seeley, T. D., and R. A. Leven. 1987. A colony of mind: The beehive as thinking machine. *The Sciences* 27: 38–43.

Silvertown, J., and G. M. Gordon. 1989. A framework for plant behavior. *Annual Review of Ecology and Systematics* 20: 349–66.

Sinnott, E. W. 1960. *Plant morphogenesis*. New York: McGraw Hill Book Company.

Slade, A. J., and M. J. Hutchings. 1987. Clonal integration and plasticity in foraging behavior in *Glechoma hederacea*. *Journal of Ecology* 75: 1023–36.

Stenhouse, D. 1974. *The evolution of intelligence: A general theory and some of its implications*. London: George Allen and Unwin.

Stocker, O. 1960. Physiological and morphological changes in plants due to water deficiency. In *Plant-water relationships in arid and semi-arid conditions*, 63–104. Geneva: U.N.E.S.C.O.

Sultan, S. E. 2000. Phenotypic plasticity for plant development, function and life history. *Trends in Plant Sciences* 5: 537–41.

Thaler, D. S. 1994. The evolution of genetic intelligence. *Science* 264: 1698–99.

Townsend, C. O. 1897. The correlation of growth under the influence of injuries. *Annals of Botany* 40: 509–32.

Trewavas, A. J. 1988. The evolution controversy: A network view. *Evolutionary Trends in Plants* 2:1–5.

———. 1992. Growth substances in context: A decade of sensitivity. *Biochemical Society Transactions* 20: 102–8.

———. 1998. The importance of individuality. In *Plant responses to environmental stresses*, ed. H. R. Loerner, 27–43. New York: Marcel Dekker.

———. 1999. Le calcium c'est la vie; Calcium makes waves. *Plant Physiology* 120: 1–6.

———. 2000. Signal perception and transduction. In *Biochemistry and molecular biology of plants*, ed. B. B. B. Buchanan, W. Gruissem, and R. L. Jones, 930–88. Bethesda, MD: American Society of Plant Physiologists.

———. 2002. Mindless mastery. *Nature* 415: 841.

———. 2003. Aspects of plant intelligence. *Annals of Botany* 92: 1–20.

———. 2004. Aspects of plant intelligence: An answer to Firn. *Annals of Botany* 93: 353–57.

Turkington, R., and E. Klein. 1991. Integration among ramets of *Trifolium repens. Canadian Journal of Botany* 69: 226–28.

Turkington, R., R. Sackville Hamilton, and C. Gliddon. 1991. Within-population variation in localised and integrated responses of *Trifolium repens* to biotically patchy environments. *Oecologia* 86: 183–92.

Verdus, M. C., M. Thellier, and C. Ripoli. 1997. Storage of environmental signals in flax: Their morphogenetic effect as enabled by a transient depletion of calcium. *The Plant Journal* 12: 1399–1410.

Vertosick, F. T. 2002. *The genius within: Discovering the intelligence of every living thing.* New York: Harcourt Inc.

Vertosick, F. T., and R. H. Kelly. 1991. The immune system as a neural network: A multi-epitope approach. *Journal of Theoretical Biology* 150: 225–37.

Von Sachs, J. 1879. *Lectures on the physiology of plants.* Trans. H. Marshall, 1887. Oxford: Oxford at the Clarendon Press.

Waddington, C. H. 1957. *The strategy of the genes.* London: Jonathan Cape.

Warden, C. J., T. N. Jenkins, and L. H. Warner. 1942. Metaphyta. In *Comparative psychology: A comprehensive treatise*, 180–286. Vol. II of *Plants and invertebrates.* New York: Ronald Press Company.

Warwick, K. 2001. *The quest for intelligence.* London: Judy Piatkus Ltd.

Wijesinghe, D. K., and M. J. Hutchings. 1999. The effects of environmental heterogeneity on the performance of *Glechoma hederacea*: The interactions between patch contrast and patch scale. *Journal of Ecology* 87: 860–72.

Wright, S. 1982. Character changes, speciation and the higher taxa. *Evolution* 36: 427–41.

Yamada, T., T. Okuda, M. Abdullah, M. Awang, and A. Furukawa. 2000. The leaf development process and its significance for reducing self-shading of a tropical pioneer tree species. *Oecologia* 125: 476–82.

Zhong, G.Y., and J. Dvorak. 1995. Chromosomal control of the tolerance of gradually and suddenly-imposed salt stress in the *Lophopyrum elongatum* and wheat genomes. *Theoretical and Applied Genetics* 90: 229–36.

6 CONVERGENT EVOLUTION, SERENDIPITY, AND INTELLIGENCE FOR THE SIMPLE MINDED

Nigel R. Franks

Is evolution a completely blind, unguided, accidental process? If so, the richness of the biological world is simply the result of countless happy accidents (i.e., serendipity) and cumulative selection. Evolution is blind, in the sense that there is no foresight: the last step is only visible with 20/20 hindsight. Much is accidental, but crucially certain accidents make other accidents more likely. As Augustus de Morgan wrote in *A Budget of Paradoxes* (1872), "Great fleas have little fleas upon their backs to bite 'em, And little fleas have lesser fleas, and so ad infinitum" (see also Jonathan Swift, "On Poetry"). One thing leads to another because nature, we are told, abhors a vacuum. An opportunity welcomes the ultimate opportunist—evolution by natural selection—the quintessential gambler. The clearest evidence that certain accidents make other accidents more likely is convergent evolution (Conway Morris 2003). From different starting points, different lineages appear to home-in upon the same syndromes. However, such evolutionary convergence is the result of accidents and selection begetting further accidents and adaptations that are foundations for further change. Convergent trends arise because certain temporary solutions are better than others; hence, certain evolutionary scenarios and syndromes are more likely than others.

This is biology's golden age with the promise of a new scientific enlightenment and a new understanding of the natural world and our lowly place herein. Yet, we have evolved the intelligence that enables us to contemplate such issues in a seemingly meaningful way. So what of the evolution of the ultimate adaptation—intelligence?

The definition I will use here is that intelligence is the ability to solve problems. There are doubtless innumerable drawbacks to such a broad definition, but I suggest that it has some advantages. The definition of intelligence needs to be broad because intelligence is multidimensional even for just one species: our own (Gould 1981). In addition, this definition avoids the arguably impossible task of prizing apart behavior and physiology.

Here, I will consider one case study in intelligence and very briefly its independently evolved analogue. These examples, from ants and bees, strongly suggest that intelligent systems can evolve from surprisingly few, small steps. They also exemplify extreme evolutionary convergence.

We will first visit an alien society—one perhaps unlike our own—in which there is intelligence at both the individual and collective level. We will make this visit to another realm, both foreign and Lilliputian, better to understand how evolution can conjure up intelligence.

To begin at the end: the answer is that intelligence can be surprisingly simple.

The devil is in the detail, so let's first unroll the natural history.

The society I have in my mind, and in my laboratory, is the diminutive rock ant *Temnothorax* (formerly *Leptothorax*) *albipennis*. These ants live, between slivers of rock, as complete colonies of one queen, several hundred workers, and the brood. They are so tiny that such a society could easily live within your wristwatch if you first tipped out all the gubbins. If their natural fragile nest is damaged, they can do little to repair it, and they frequently move house, lock, stock and barrel. Their natural nests are usually almost completely flat and can be mimicked in the laboratory with a nest in the form of a microscope-slide sandwich (Figure 1a and b).

These ants are brilliant house hunters, surveyors, real-estate agents, and removal companies (Franks et al. 2002; 2003a; 2003b). They can measure many attributes of potential nest sites, and they use one of the most thorough and sophisticated decision-making strategies to weight

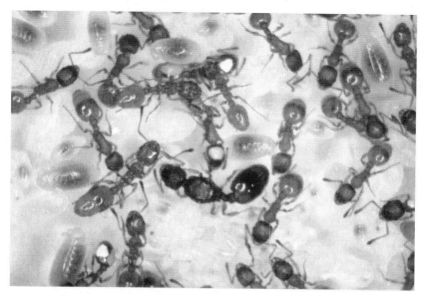

Figure 1a. A colony of the ant *Temnothorax albipennis* housed in a nest made of microscope slides. The single large queen is at the center. Each worker ant is between 2mm and 3mm long. © Nigel R. Franks

Figure 1b. A close-up of two *Temnothorax albipennis* workers. One is individually marked with paint. © Nigel R. Franks

and sum the attributes of nests (Franks et al. 2003b). They can measure the floor area of a potential nest, working entirely alone, in total darkness; vagaries in nest shape do not distract them (Mallon and Franks 2000; Mugford et al. 2001). They also take into account the amount of head room in the cavity; light ingress and the abundance and width of nest entrances (Franks et al. 2006a); issues of nest hygiene (Franks et al. 2005); and the proximity of nasty neighbors (Franks et al. 2007a). They can encode their evaluation of the nest site, allowing more or less time over the decision to allow other nests to be discovered or to rush, if needs be (Mallon et al. 2001; Franks et al. 2003a). They can share and collate their different opinions and rapidly achieve consensus through quorum sensing (Pratt et al. 2002), and they can swiftly evacuate their old nest, taking special care of the queen, and deliver everyone safely to the newly chosen home (Franks and Sendova-Franks 2000). They are magnificent!

Consider the Floor-Area Problem

How can an ant measure a large area accurately, irrespective of shape, working totally in the dark? Simple: it visits the prospective nest, lays an individual specific pheromone trail for a set amount of time, leaves, and, on return, measures the frequency at which it crosses it previous path (Mallon and Franks 2000; Mugford et al. 2001). We call this Buffon's needle algorithm. Comte George Buffon in 1777 proposed a method of estimating π empirically by counting the number of times a needle, dropped randomly, crossed parallel straight lines inscribed on the surface of a plane. Buffon's formulation can easily be rearranged to estimate area from the frequency at which two sets of lines of known length cross. This is not to say that our ants need to know π or how to do a complex calculation; rather, Buffon's mathematics shows that area is inversely proportional to the frequency at which the two sets of lines cross. It is all rather simple.

Furthermore, colonies of all sizes have the same preference for nest area (Franks et al. 2006b), even though the dynamics of their decision making may differ (Dornhaus and Franks 2006). They all prefer nests that would perfectly suit fully grown colonies (Franks et al. 2006b). So, they are either choosing one that is right for their fully gown colony now or choosing something to grow into.

Consider the Ideal Consumer and the Weighted Additive Strategy

But the ants do not just care about the floor area of a potential cavity; they also assess headroom, entrance width, and darkness. Holding floor area constant, we gave emigrating ant colonies binary choices between nests that varied in these traits. Twelve nest designs and more than 340 colony emigrations later, we can conclude that the ants prefer dark nests most of all; next, they then favor ones with good headroom; and last, they like narrow entrances rather than wide ones. Such is the ranking of their *desiderata*. In sum, the ants show consistent preferences, consistent rankings, and transitivity (if they prefer A to B and B to C, they prefer A to C). Thus, they are logical and rational (again, in contrast to us). But what surprised us most was that two lower-ranking traits in combination could *outweigh* a single higher-ranking variable. Thus, although in general the ants prefer dark nests, given the choice between a dark nest, with poor headroom and too wide an entrance and a too-bright nest with good headroom and a desirably narrow entrance, the ants consistently choose the latter (see Figure 2). This is convincing evidence that the ants take all variables into account and somehow weigh them up to make their final choice. Indeed, we have argued that the ants almost certainly exhibit the "weighted addi-

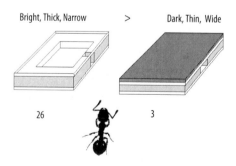

Bright, Thick, Narrow > Dark, Thin, Wide

26 3

Figure 2. Given the choice, twenty-six out of twenty-nine colonies selected a bright, thin, narrow-entranced nest to one that was dark, thin, and wide-entranced. Other experiments had shown that, all else being equal, the ants like dark nests, those that are tall (i.e., have good head room), and those with a narrow (easily defended) entrance. This experiment shows that two desirable attributes (tall and narrow) can together outweigh the single most desirable attribute of a nest. This is strong evidence that the ants are using the most sophisticated consumer decision-making system—the weighted additive strategy.

tive strategy," one of the most difficult but thorough decision-making schemes (Franks et al. 2003b). How they do this we do not yet know. We do know, however, that it can be done by lone individuals (i.e., not by mass action). In three emigrations by three different colonies, 92% of the thirty-eight ants that visited both nests before beginning to recruit to either of them initiated recruitment only to the superior one (Pratt et al. 2002; Franks et al. 2002). This is strong evidence for direct comparison by individuals and, in turn, that individuals can accomplish weighted additive decision making. How such tiny-brained individuals (each probably with less than 100,000 neurons) can collate such different sensory inputs (light, assessed through their eyes; nest height and entrance width measured by proprioceptors; area by trail-crossing frequencies picked up with their antennae, etc.) into a single score is not known. This issue is known as the "binding problem" (Roskies 1999). That individual ants, 2 mm long, find this neither a bind nor a problem might tell us much about our own highbrow intelligence of which we are so proud. For these house-hunting ants, the binding problem would simply require the collation of different sensory inputs. At its simplest, this would only require their brains to have the neuronal "wiring" to sum different inputs. In theory, this could also be rather simple. Intriguingly, their individual brains might need to do internally a form of quorum sensing, similar in principle to the quorum sensing they do externally as a committee of individuals as servants of their society (see later).

Consider the Simplest Way To Encode Enthusiasm

How do individual ants communicate their quantitative assessment of the quality of a nest to their colony mates? Simple: they hesitate over poor nests and swiftly vote with their feet for good ones (Mallon et al. 2001). If they like a nest, they start recruiting to it, by tandem running, sooner rather than later. If they are not smitten by a nest, they are prone to dilly and dally. Thus, nest quality is encoded in time-dependent enthusiasm. The worse the nest, the more they procrastinate.

Figure 3. Tandem running. The ant at the front has found a suitable nest and leads its nest mate there, teaching it the route. © Stephen Pratt

Consider Teaching

How does one ant solicit a second opinion? First, by actively teaching a naïve ant the route to the prospective new nest (Franks and Richardson 2006). They do this by tandem running (see Figure 3). Indeed, tandem running in these ants was the first behavior to qualify as formal teaching in any nonhuman animal (Franks and Richardson 2006). Such tandem running is slow, but one ant has shown another the way to the new nest, and both leader and disciple may, in turn, show others the way. Thus, the number of ants that both know the way to, and the quality of, a new nest may slowly snowball.

Consider the Fireperson's Lift

As all academics know, teaching by example can be excruciatingly slow. Is there not a swifter way to take the student to the right answer? In this case, the answer is yes. One ant simply picks up another and runs off with it at top speed to the new nest site. Indeed, such piggyback carrying is almost exactly three times faster than tandem running, and, once it

Figure 4. After a tandem-running ant has found a sufficient number (the quorum threshold) of its nest mates in the new nest, it switches to carrying behavior, and the emigration proceeds at full speed. The collective decision has been made. © Stephen Pratt

has begun, it is usually clear which nest the ants have chosen (Figure 4). So the transition from recruitment by tandem running to recruitment by carrying seems to coincide with a key decision point. So how does an ant know when to switch from tandem running to carrying? The answer is: they vote.

Consider the Vote

The ants exhibit quorum sensing (Pratt et al. 2002). Under standard laboratory conditions, they switch from tandem running only when they have encountered about twelve of their nest mates in the new nest site (Figure 5 a, b, and c). This is a sufficiently high number that an individual ant is unlikely to have led so many of its nest mates there personally. Indeed, it is likely that such a number can only be raised by several nest mates recruiting others to the nest site. So, the quorum indicates that several ants share the view that this nest site is suitable. Quorum sensing is a voting and opinion-polling procedure. It translates individual decision making into collective decision making. By controlling access to a new nest, we can demonstrate the ants' use of quorum sensing (Pratt et al. 2002).

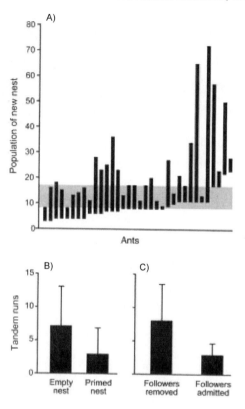

Figure 5. (a) Recruiting ants use tandem runs when the population of their nest mates at a new nest site is small, and they use transport when it is high. The "population of new nest" is the estimate of the new nest-site population at which ants switch from tandem running to transport. The ends of each black bar show the upper and lower bounds of the estimate for a single recruiter. (b) When recruits are prevented from entering a new nest, the number of tandem runs is significantly higher than if it is primed with nest mates. (c) Tandem leaders do significantly more recruiting if their followers are removed than if such followers are allowed to enter the nest and contribute to the quorum. Figures 5 a, b and c are from Franks et al. 2002; original study by Pratt et al. (2002).

Consider Adaptive Procrastination

What is remarkable about this system is that it is neither the fastest imaginable nor, indeed, the fastest that can be demonstrated experimentally (Franks et al. 2003a), but it may be a very good system to find the best-of-N available nests at reasonable speed. In other words, the ants are not just good at binary choices, but they can pick the best available nest in a large array of mediocre ones. By building time lags (procrastination) into their decision making, the ants give themselves time to discover the best available nests (Planqué et al. 2006).

Consider Speed Versus Accuracy

But what if speed is of the essence? Would you be calm and placid if you could smell formic acid? When we rudely invite our ants to emigrate in harsh conditions (either a howling gale or in the presence of a sinister whiff of formic acid, as might be released by their enemies, other bigger ants), individual ants do not consult their colleagues but rapidly make up their own "minds." Quorum sensing disappears, and the emigration proceeds more quickly, but not without cost. Decide at speed and repent at leisure. When they make quicker decisions, the ants are more error-prone. In harsh conditions, the ants are more likely to start transferring their brood to the lesser of two available nest sites (Franks et al. 2003a). For detailed mathematical models of this decision-making system, see Pratt et al. (2005) and Marshall et al. (2006).

Consider the Property Ladder

These ants can also organize emigrations from old nests, even if they are still fully intact and habitable. They only do so if the new property is a sufficient upgrade from the old one (Dornhaus et al. 2004).

Consider the Monarch

What else does the ant colony have to worry about? That is, what else has natural selection focused upon? Well, the queen mother, of course! When, for maximum security, should the queen be taken to the new nest? Her greatest protection is in the blanket of workers around her. So she should be transferred quickly to the new nest when half of the workforce is in the new nest (and half is in the old). This is exactly what the ants do (Franks and Sendova-Franks 2000). How do they know? As yet we do not know— but we will go to work on this soon. My bet is that the mechanism will be simple, fault-tolerant, and robust.

Consider Planning for the Future

Animal learning is often thought to occur only for immediate rewards, such as food. However, we show that these ant colonies (*Temnothorax albipennis*) learn about the housing stock in their neighborhood even when they do not need to emigrate. Then, when they do need to find a new home, they discriminate against low-quality nests that they have previously marked with scents and remembered by learning landmarks. In this way, they are able to focus their search elsewhere for better new nest sites. In effect, these behaviors allow the ants to plan for the future (Franks et al. 2007b).

Recently, Raby et al. (2007) have shown that western scrub-jays (*Aphelocoma californica*) also plan for the future by storing particular foods where they will be needed most. This might even involve "mental time travel" because it seems to involve (1) novel behavior based on learning and (2) the animal's anticipating a different motivational state to its current one. However, our findings suggest that ants may also, in effect, anticipate the future through individuals acquiring private information by learning landmarks and providing public information in the form of pheromones. Moreover, the motivational states of the different sets of ants depositing pheromones, when the colony does not need to emigrate, and those reacting to them, when the colony does need a new home, are likely to be different. So, planning for the future can be a social activity based on relatively simple rules without any form of mental time travel.

Consider the Learned Society

Recently, Langridge et al. (2004) have shown that these ant societies become more efficient at emigrating when they repeat the task of house hunting. This raises the possibility that these superorganisms can learn.

Consider Here How Simple It Was for Evolution to Meld Individual and Collective Intelligence

Well, I hope that you have joined me in enjoying the extraordinary abilities of these ants. Some parts of the story are yet to be fully worked out: how individual ants make weighted additive decisions, and how they know when to move the queen. Where, however, we have worked out the mechanisms of this brilliant decision-making system, they are surprisingly and very satisfyingly simple. What might this tell us about the evolution of intelligence? Here we have a flexible highly adaptive information-gathering and decision-making system, that is, a form of intelligence, yet (almost) every step is small. Here this evolutionary "Mount Improbable" (see, for example, Dawkins 1996)—one towering adaptive peak among an ever-changing seascape of possibilities (see, for example, Wilson 1982)—seems to have been conquered one small, easily negotiated, step at a time.

Consider the following scenario. In the beginning was the ant—a solitary decision maker. She simply started to carry her nest mates to any reasonable nest she discovered. But selection punished such "hares" and favored colonies with slower more deliberate workers who took their time and their friends in tandem—and, thus, just by chance, took into account the opinions of their nest mates. But these careful plodding tortoise-like forms can also step on the accelerator when they sense the need.

Consider Honey Bees

Remarkably, honeybees when they swarm have an almost identical set of house-hunting problems to those of the *Temnothorax* ants I have described (see Seeley 1977, 1982; Seeley and Buhrman 1999, 2001; Seeley and Tautz 2001; Seeley and Visscher 2004; Seeley et al. 2006). The honeybees also weigh up a whole set of different attributes of potential new hives and may also use quorum sensing (for a recent comparative review of the ants and the honeybees, see Visscher 2007). Although many of the mechanisms differ in detail, at the deepest logical level, the decision-making systems of these two very different insect societies are extremely similar (Franks et al.

2002, Britton et al. 2002, Franks and Dornhaus 2003). This is a remarkable example of convergent evolution in decision making and intelligence.

Consider the Ant

My guess is that, when we can dissect, reconstitute, and determine (qualitatively and quantitatively) how the parts make the whole for other intelligent systems, such as brains, we will see that these intelligent systems are also formed from surprisingly simple elements. Thus, these seemingly improbable adaptations may also be not so difficult for evolution by natural selection to construct.

Of course, the intelligence of ants (and most other animals) may be quite different in character to the higher intelligence of humans. It is algorithmic rather than intellectual. In other words, it is based on rules of thumb and recipe-like procedures. Nevertheless, I suspect that much of the intelligence of human beings will also be shown to be based on algorithms composed of if-then-else rules and the like.

When we marvel at what is intelligent behavior in ants, what we are also really glimpsing is the magnificently thorough way in which evolution has created and explored diversity and found comparatively simple solutions to complex problems.

I strongly suspect that, because the problem-solving abilities of ants (and other animals) are algorithmic rather than intellectual, we might be able to derive useful insights from studying their behavior.

Why do I study ants? Because they often solve problems partly at a collective colony level, but unlike brains, we can easily observe the behavior of the parts and the whole and we can take the entire colony apart and put it back together again.

So, perhaps, new lessons will be learned not from our obsession with our own intelligence but by actually taking King Solomon's advice to "Go to the ant . . . consider her ways and be wise" (Prov. 6). It is an immense privilege to have the opportunity to study these alien societies, and all the evidence suggests that we will be wise to learn from these frequently overlooked organisms. They might well be the key to understanding the evolution of complexity and intelligence.

Conclusion

The decision-making systems of the insect societies, considered here, are intelligent. They involve information gathering, evaluation, deliberation, consensus building, choice, and implementation, and they are sensitive to context. Furthermore, it seems eminently plausible that these and other intelligent systems have evolved from a few, small, simple steps, each easily negotiated by chance and selection with the outcome captured in the genotype and expressed in the extended phenotype.

So, if the ultimate adaptation—intelligence—can be explained as the product of unadulterated natural selection, surely it is misleading to suppose that there is purpose in evolution.

Rigorous science proceeds through the experimental falsification of hypotheses. It is by necessity entirely naturalistic (Ruse 2003; Hull 2004). Being entirely naturalistic, science has nothing to say directly about worldviews that postulate the supernatural. However, this also means that worldviews that are based on beliefs in the supernatural have nothing to contribute to science.

Science does make progress, even though the endeavor is endless. As science explains more, it reveals much more to be explained. Scientific progress is exemplified in associated technological achievements. For innumerable examples, one need look no further than to medicine, engineering, and informatics. Indeed, science is arguably the one activity of humanity with demonstrable long-term progress. Science leaves less and less to be explained in supernatural terms. This does not lead inevitably to a barren and bleak worldview. On the contrary, science should help to liberate us all from myth, superstition, and sophistry. Modern biology would seem to suggest that evolution has no purpose: in other words, no immortal hand or eye framed our fearful destiny (with apologies to William Blake). If such is the case, one distinct possibility looms: is it solely our responsibility to determine our own purpose and destiny?

Acknowledgments

The studies I have referred to are all either published or in press. I thank all of my co-authors but especially Ana Sendova-Franks (especially for the discussion of the definition of intelligence adopted in this chapter), Eamonn Mallon, Stephen Pratt, David Sumpter, Nick Britton, Tom Seeley, Anna Dornhaus, Elizabeth Langridge, Tom Mischler, Matthew Hamilton, Helen Bray, Martin Stevens, Jon Fitzsimmons, Ross Hawkins, and Harriet Shere. I also thank every other member of my lab, especially our beloved ants. Not one of these individuals is responsible for the heretical views expressed in this paper. That responsibility is mine alone. I also thank the Templeton Foundation and Simon Conway Morris for their invitation to participate in a workshop with a wealth of fascinating discussions. Their hospitality is all the more remarkable given the extreme divergence in certain of our views.

Figures 1a and 1b were photographed by Nigel Franks. Figures 3 and 4 were photographed by Stephen Pratt. Figure 5 was published originally in the essay "Information flow, opinion polling, and collective intelligence in house-hunting social insects" in *Philosophical Transactions of the Royal Society of London Biological Sciences* 375: 1567–83.

References

Britton, N. F., N. R. Franks, S. C. Pratt, and T. D. Seeley. 2002. Deciding on a new home: How do honeybees agree? *Proceedings of the Royal Society of London: Biological Sciences* 269: 1383–88.

Conway Morris, S. 2003. *Life's solution: Inevitable humans in a lonely universe.* Cambridge: Cambridge University Press.

Dawkins, R. 1996. *Climbing mount improbable.* New York: W.W. Norton & Company.

Dornhaus, A., and N. R. Franks. 2006. Colony size affects collective decision-making in the ant *Temnothorax albipennis. Insectes Sociaux* 53: 420–27. (doi: 10.1007/s00040-006-0887-4).

Dornhaus, A., N. R. Franks, R. M. Hawkins, and H. N. S. Shere. 2004. Ants move to improve: Colonies of *Leptothorax albipennis* emigrate whenever they find a superior nest site. *Animal Behaviour* 67:959–63.

Franks, N. R., and A. Dornhaus. 2003. How might individual honeybees measure massive volumes? *Biology Letters. Proceedings of the Royal Society of London: Biological Sciences Supplement* 270: S181–S182.

Franks, N. R., and A. B. Sendova-Franks. 2000. Queen transport during ant colony emigration: A group-level adaptive behaviour. *Behavioral Ecology* 11: 315–18.

Franks, N. R., and T. Richardson. 2006. Teaching in tandem-running ants. *Nature* 439: 153 (doi: 10.1038/439153a).

Franks, N.R., S. C. Pratt, E. B. Mallon, N. F. Britton, and D. J. T. Sumpter. 2002. Information flow, opinion-polling and collective intelligence in house-hunting *Social Insects Philosophical Transactions of the Royal Society of London: Biological Sciences* 357: 1567–83.

Franks, N. R., A. Dornhaus, J. P. Fitzsimmons, and M. Stevens. 2003a. Speed versus accuracy in collective decision-making. *Proceedings of the Royal Society of London: Biological Sciences* 270: 2457–63.

Franks, N. R., E. B. Mallon, H. E. Bray, M. J. Hamilton, and T. C. Mischler. 2003b. Strategies for choosing among alternatives with different attributes: Exemplified by house-hunting ants. *Animal Behaviour* 65: 215–23.

Franks, N. R., J. Hooper, C. Webb, and A. Dornhaus. 2005. Tomb evaders: House-hunting hygiene in ants. *Biology Letters* 1: 190–92.

Franks, N. R., A. Dornhaus, B. G. Metherell, T. R. Nelson, S. A. J. Lanfear, and W. S. Symes. 2006a. Not everything that counts can be counted: Ants use multiple metrics for a single nest trait. *Proceedings of the Royal Society B* 273: 165–69.

Franks, N. R., A. Dornhaus, C. S. Best, and E. L. Jones. 2006b. Decision-making by small and large house hunting ant colonies: One size fits all. *Animal Behaviour* 72: 611–16 (doi: 10.1016/j.anbehav.2005.11.019).

Franks, N. R., A. Dornhaus, G. Hitchcock, R. Guillem, J. Hooper, and C. Webb. 2007a. Avoidance of conspecific colonies during nest choice by ants. *Animal Behaviour* 73: 525–34 (doi:10.1016/j.anbehav.2006.05.020).

Franks, N. R., J. W. Hooper, A. Dornhaus, P. J. Aukett, A. L. Hayward, and S.M. Berghoff. 2007b. Reconnaissance and latent learning in ants. *Proceedings of the Royal Society B* 274: 1505–9 (doi: 10.1098/rspb.2007.0138).

Gould, S. J. 1981. *The mismeasure of man.* New York: W.W. Norton and Co.

Hull, D. L. 2004. Complexity, design and natural selection. *BioScience* 54:162–64. (Book review of Michael Ruse, *Darwin and Design: Does Evolution have a Purpose?* [Cambridge, MA: Harvard University Press].)

Langridge, E. A., N. R. Franks, and A. B. Sendova-Franks. 2004. Improvement in collective performance with experience in ants. *Behavioural Ecology and Sociobiology* 56: 523–29.

Mallon, E., and N. R. Franks. 2000. Ants estimate area using Buffon's needle. *Proceedings of the Royal Society of London: Biological Sciences* 267: 765–70.

Mallon, E. B., S. C. Pratt, and N. R. Franks. 2001. Individual and collective decision-making during nest site selection by the ant *Leptothorax albipennis. Behavioral Ecology and Sociobiology* 50: 352–59.

Marshall, J. A. R., A. Dornhaus, N. R. Franks, and T. Kovacs 2006. Noise, cost and speed-accuracy trade-offs: Decision-making in a decentralized system. *Journal of the Royal Society: Interface* 3: 243–54 (doi: 10.1098/rsif.2005.0075).

Mugford, S. T., E. B. Mallon, and N. R. Franks. 2001. The accuracy of Buffon's needle: A rule of thumb used by ants to estimate area. *Behavioral Ecology* 12: 655–58.

Planqué, R., A. Dornhaus, N. R. Franks, T. Kovacs, and J. A. R. Marshall. 2006. Weighting waiting in collective decision-making. *Behavioral Ecology and Sociobiology* 61: 347–56 (doi: 10.1007/s00265-006-0263-4).

Pratt, S. C., E. B. Mallon, D. J. T. Sumpter, and N. R. Franks. 2002. Quorum sensing, recruitment, and collective decision-making during colony emigration by the ant *Leptothorax albipennis. Behavioural Ecology and Sociobiology* 52: 117–27.

Pratt, S. C., D. J. T. Sumpter, E. B. Mallon, and N. R. Franks. 2005. An agent-based model of collective nest choice by the ant *Temnothorax albipennis*. *Animal Behaviour* 70: 1023–36.

Raby, C. R., D. M. Alexis, A. Dickson, and N. S. Clayton. 2007. Planning for the future by western scrub-jays. *Nature* 445: 919–21 (doi:10.1038/nature005575).

Roskies, A. L. 1999. The binding problem. *Neuron* 24: 7–9.

Ruse, M. 2003. *Darwin and design: Does evolution have a purpose?* Cambridge, MA: Harvard University Press.

Seeley, T. 1977. Measurement of nest cavity volume by the honeybee (*Apis mellifera*). *Behavioural Ecology and Sociobiology* 2: 201–27.

———. 1982. How honey-bees find a home. *Scientific American* 247: 158–68.

Seeley, T. D., and S. Buhrman. 1999. Group decision making in swarms of honeybees. *Behavioural Ecology and Sociobiology* 45: 19–31.

———. 2001. Nest-site selection in honeybees: How well do swarms implement the "best-of-N" decision rule? *Behavioural Ecology and Sociobiology* 49: 416–27.

Seeley, T. D., and J. Tautz. 2001. Worker piping in honey bee swarms and its role in preparing for liftoff. *Journal of Comparative Physiology A* 187: 667–76.

Seeley, T. D., and P. K. Visscher. 2004. Quorum sensing during nest-site selection by honeybee swarms. *Behavioural Ecology and Sociobiology* 56: 594–601.

Seeley, T. D., P. K. Visscher, and K. M. Passino. 2006. Group decision making in honey bee swarms. *American Scientist* 94: 220–29.

Visscher, P. K. 2007. Group decision making in nest-site selection among social insects. *Annual Review of Entomology* 52: 255–75 (doi10.1146/annurev.ento.51.110104.151025).

Wilson, E. O. 1982. Of insects and man. In *The biology of social insects*, ed. M. D. Breed, C. D. Michener, and H. E. Evans, 1–3. Boulder, CO: Westview Press.

7 CANNY CORVIDS
AND POLITICAL PRIMATES

A Case for Convergent Evolution in Intelligence

Nicola S. Clayton

Nathan J. Emery

If men had wings and bore black feathers,
few of them would be clever enough to be crows.

Reverend Henry Ward Beecher

In the latter half of the nineteenth century, Charles Darwin suggested that mental characteristics are subject to natural selection in much the same way as morphological traits, and, thus, we would expect some characteristics of human intelligence to be present in other descendants of our primate lineage (Darwin 1872). By mental characteristics, we mean more than just the ability to learn and remember. For the purposes of this chapter, intelligence refers to the ability to think, reason, and solve novel problems. Specifically, intelligent beings can think not only about the here-and-now, but they can also reminisce about their past and plan for their future (so-called mental time travel). They can also think about what others might be thinking and how this might be different to what they themselves think (theory of mind). Furthermore, intelligent beings should be capable of devising novel solutions to problems, such as the manufacture of special tools to acquire otherwise unobtainable foods.

Presumably, the development of these mental characteristics confers some reproductive advantage, especially for long-lived animals that require a sophisticated appreciation of their physical and social world in order to survive the trials and tribulations of life. Indeed, a number of hypotheses has been proposed to account for the enhanced intellectual capacities of primates, and these broadly fall into two categories: physical and social. Milton (1981) has argued one physical challenge that primates face is to monitor the availability of fruits and other widely dispersed, ephemeral, high-quality foods; and to do this efficiently, they should remember which foods are where and how ripe they are now, in order to predict when they will be ripe. In addition to spatiotemporal mapping, there may other physical challenges associated with foraging, particularly extractive foraging, which may require tools to be manufactured and used for such purposes (Parker and Gibson 1977; Byrne 1997).

However, Jolly (1966) and Humphrey (1976) independently proposed an alternative hypothesis for the evolution of primate intelligence, namely, that it is the ability to survive the political dynamics of a complex social world that has been the primary driving force shaping primate intelligence. This "social function of intellect" hypothesis states that the complexities of social life have led to an increase in general intelligence, and Dunbar (1992) has further suggested this also leads to a dramatic increase in the relative size of the neocortex during primate evolution. It is certainly plausible to argue that surviving the trials and tribulations of a complex social world makes intellectual demands on many primates. Individuals need to know who is who, they need to keep track of who did what to whom, where and when, and to use this information to predict the actions and intentions of other individuals in their social network, as well as understanding how these relationships change over time (Barrett et al 2003). In short, the need for effective competition and cooperation with conspecifics may have provided the main selective advantage for the evolution of primate intelligence (Byrne and Whiten 1988; Dunbar 1998).

That said, there is no reason to assume that intelligence is restricted to primates or that such abilities have evolved only once. Indeed, we shall argue that there is good reason to believe that complex mental characteristics have evolved several times and that the existence of intelligence in different, distantly related lineages must have arisen as a result of convergent evolution in species facing similar social and physical problems. By definition, *convergence* refers to similarities between groups that arise as a result of adaptation to similar selection pressures, not because of phylogenetic relatedness, and the more distantly related the two groups, then the stronger the case for convergence. As Conway Morris (2003) has argued, there are many examples of morphological traits that have evolved multiple times in distantly related lineages. One of the best examples of such evolutionary convergence is the development of the camera eye, which has evolved *de novo* three times: namely, in the vertebrates, in some cephalopods (squid and octopus), and also in one group of marine annelid worms (the alciopids).

With regard to the evolution of intelligence, Marino (2002) has made a convincing case for the convergent evolution of intelligence in the cetaceans (dolphins, whales, and porpoises) and the anthropoid primates (monkeys, apes, and humans). There may be other groups of mammals, such as elephants, that also share these mental characteristics. The fact that these abilities are not found in lineages that are more closely related to the primates—for cetaceans and primates diverged at least sixty-five million years ago—suggests that complex cognition has evolved within the mammals more than once. But perhaps the most dramatic case for convergent evolution of cognition comes from comparing primate cognitive abilities with those of crows, given that the common ancestor of mammals and birds lived over 280 million years ago and that not all birds and mammals share the complex mental abilities found in crows and primates. Indeed, "birds as a whole are a rich source of insights into the prevalence of evolutionary convergence, as well as having some striking similarities with other groups" (Conway Morris 2003, 138).

Why Study Intelligence in Crows?

If one were looking for avian candidates of intelligence, folklore would point toward two groups, the parrots and the corvids (the crow family). Humans have been intrigued by the mental abilities of several members of the corvids, which includes jays and ravens, as well as crows. Stories of ravens, for example, go back long before the sacred texts of Christianity (Sax 2003). Anecdotes abound, and folklore is rich in examples—from the arms of the Baron von Rindscheit symbolizing the union between the strength of the boar and the wisdom of the crow to the series of Aesop's fables about the canny cleverness of crows. The nature writer David Quammen (1985, 30) claimed that each member of the crow family is "so full of prodigious and quirky behaviour that it cries out for interpretation not by an ornithologist but by a psychiatrist." His theory is that the crows are bored and constantly up to mischief, too clever for their own good, like very bright children!

There are a number of scientific reasons for believing that these animals are very intelligent. Like primates, corvids are particularly good at solving laboratory tasks that rely on the ability to abstract a general rule to solve the task and transfer the general rule to new tasks, whereas pigeons show no evidence of abstraction and instead rely on simple rote learning (Wilson et al. 1985; Mackintosh 1988). Unfortunately, parrots have not been tested on these tasks, so we do not know whether parrots are also capable of abstraction.

Another similarity is that both the primates and the crows and parrots have very large brains relative to body size (Emery and Clayton 2004a). Although there is some variation in relative brain size between different crow species (Voronov et al. 1994), they all have very large brains relative to all other families of birds (Rehkamper et al. 1991). This is also the case for parrots, and some highly social species such as the African grey and various macaws have very large brains relative to body size (Iwaniuk et al. 2005; see also Portmann 1947; Burish et al. 2004). Figure 1 shows that the relative forebrain size of corvids and parrots is as large as that of the nonhuman apes.

It is important to note that the structural organization of the brains of birds and mammals is very different and that they evolved from different reptilian ancestors. For example, avian brains have a nuclear structure, whereas mammalian ones have a laminar arrangement (see Emery and Clayton 2005 for a recent review). In terms of the neural bases for

Figure 1. A regression of log brain weight and log body weight for various species of corvid, parrot, and ape. The data for birds were obtained from Iwaniuk and Nelson (2003), while the ape data were taken from Rilling and Insel (1999).

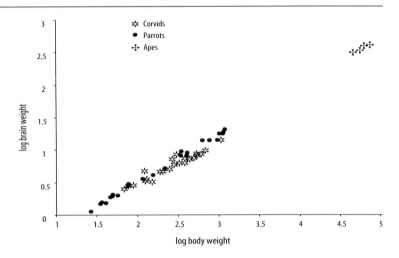

intelligence, one of the key differences is that birds do not have a cortex, whereas mammals do. In both human and nonhuman primates, it is one specific region of the cortex, namely, the prefrontal cortex, that is thought to play a critical role in thinking. Although birds do not have a prefrontal cortex, the nidopallium caudolaterale (formerly, the neostriatum cadolaterale; Reiner et al. 2004) appears to be functionally equivalent (Mogensen and Divac 1982; Reiner 1986), and the volume of this brain region correlates with some measures of intelligence such as tool use (Lefebvre et al. 2002) and innovation rate as measured by reported frequencies of novel behavior patterns (Lefebvre et al. 1997). Furthermore, corvids have the largest neostriatum, relative to overall brain and body size, of any group of birds (Emery and Clayton 2004a). This large expansion of the crow neostriatum mirrors the increase in size of the prefrontal cortex in great apes (Semendeferi et al. 2002).

Another feature crows have in common with primates is that they are long lived, with an extensive developmental period in which they are dependent on their parents, which allows them ample opportunities to learn various essential skills for later life (Iwaniuk and Nelson 2003; Clayton and Emery 2007). Many species of the corvid family also live in complex social groups. For example, in the cooperatively breeding Florida scrub-jay, several closely related family members share the responsibility of raising the young with the parents. Furthermore, rooks congregate in large colonies, where juveniles associate with many nonrelatives as well as kin. In both cases, however, this long developmental period provides increased opportunities for learning from many different group members (Emery et al. 2007).

For all these reasons, Emery and Clayton (2004b) have argued that the social complexity of some crows is comparable to that of chimpanzees and that these two very distantly related families face similar challenges. Following Emery and Clayton (2004b), we shall argue that some members of the crow family possess intellectual abilities that are not only similar to some primates but are on a par with the great apes.

Evidence for Convergent Evolution of Intelligence in Crows and Primates

One feature of human intelligence is the ability to reminisce about the past (episodic memory) and plan for the future. Suddendorf and Corbalis (1997, 2007) have argued that such mental time travel is unique to humans, and, thus, animals are incapable of mentally traveling backwards in time to recollect specific past events about what happened where and when or forward to anticipate future needs. However, recent experiments in crows question this assumption by showing that one species of crow, the Western scrub-jay, can recall previous caching episodes. By *caching*, we mean that these birds hide food for future consumption and rely on memory to recover their hidden caches of food at a later date. In a series of experiments, we have shown that these birds form integrated memories of what they cached and where and when they hid it (Clayton and Dickinson 1998; Clayton, Yu, and Dickinson 2003) and that they can also keep track of who was watching when they hid particular caches and return to protect those caches appropriately at a later date (Dally et al. 2006). The jays are also capable of prospective cognition, adjusting their caching behavior in anticipation of future needs as opposed to current ones (Clayton, Yu, and Dickinson 2003; Clayton et al. 2005; Correia et al. 2007; de

Kort et al. 2007; Raby et al. 2007). The ability to remember the "what-where-and-when" of a particular episode has not yet been demonstrated in nonhuman primates. The "when" component of these personal past experiences is critical. Although multiple events can occur at the same time, you can experience only one at any given moment in time. In short, you may recall two episodes that share the same "where" or "what," but they will not share the same "when" (Clayton, Bussey, and Dickinson 2003).

A second feature of intelligence is the ability to understand and reason about the minds of other individuals and, thus, to think about what others might be thinking (theory of mind). There has been much debate about the question of whether any animal has theory of mind, in part because humans rely on language to assess these sorts of abilities. In humans, it has been suggested that the most unequivocal evidence for theory of mind lies in demonstrating that the subject can understand that another individual may have different beliefs about the world. An individual that had theory of mind could practice *tactical deception*, the intentional manipulation of another's beliefs leading to him or her to think something contrary to the truth (Byrne and Whiten 1988). The trouble with any apparent demonstration is that it is difficult to establish that the deceiver is not simply attempting to manipulate another individual's behaviour rather than his or her beliefs. A second property of theory of mind is *experience projection*, the ability to use your own experience to predict another individual's future behavior, in relation to your own. This ability has been tested only once in animals so far: in scrub-jays, not apes.

In a series of experiments, we tested whether the birds could adjust their caching strategies to minimize potential stealing by other birds, for example, by moving the food to new hiding places when other birds were not watching (Emery and Clayton 2001). Scrub-jays that had prior experience of stealing another bird's caches did move the food to new hiding places, but only if they had been observed by another bird at the time of caching and were then given the opportunity to recover and recache their food in private. If they had hidden their caches in private, however, they did not recache the food in new places when given the chance to recover them in private. One important point is that recaching is not dependent on the presence of the potential thief because the birds are always alone (in private) at the time of recovery. In order to know whether to recache, the bird must remember whether another bird was present at the time of caching. The dramatic finding was that this behavior depended on prior experience of being a thief. Jays without this experience of stealing another bird's caches did not move the caches to new places, even though they had watched other jays caching food. These results suggest that the ability to move the caches and rehide them in new sites unbeknown to the observer depends upon the previous experience of having stolen food cached by other jays, as well as on remembering whether another bird was watching them cache the food in the first place. The inference is that scrub-jays can remember not only the social context of caching (presence or absence of another bird) but can also relate information about their previous experience as a thief to the possibility of future stealing by another bird to modify their caching strategy accordingly. Other experiments on the cache protection tactics of both the scrub-jays and another fellow corvid, the raven, suggest that these birds have a complex understanding of social cognition (e.g., Bugnyar and Kotrschal 2002; Bugnyar and Heinrich 2005; Dally et al. 2004, 2005, 2006).

Another classic feature of intelligence is problem solving. Indeed, we argued that intelligent beings should be capable of devising novel solutions to problems and that one of the most dramatic examples of this is the manufacture of special tools to acquire otherwise unobtainable foods. By tool use, we mean "the external deployment of an unattached environmental object to alter more efficiently the form, position or condition of another object" (Beck 1980, 10), and this is differentiated from tool manufacture, which refers to "any modification of an object by the user or conspecific so that the object serves more effectively as a tool" (Beck 1980, 11).

The New Caledonian crow is extraordinarily skilled at making and using tools. These birds make different types of tool that have different functions (Hunt 1996). Some tools are made from *Pandanus* leaves, and these stepped-cut tools are used for probing for prey under leaf detritus. They also make hooked twig tools for poking insect larvae out of tree holes. The same tool may be used in different ways for different jobs; for example, when using the stepped-cut tools, crows make rapid back-and-forth movements for prey under soil, yet they use slow deliberate movements to spear the prey onto sharpened barbs of the leaf when the prey is in a hole (Hunt 2002). Furthermore, crows from different geographical areas have different designs of tool (Hunt and Gray 2003). The only other animals that display this diversity and flexibility in tool use and manufacture are the great apes. Thus, chimpanzees have been observed to manufacture a range of different tools that are used for specific purposes (Beck 1980), and different geographical populations of chimpanzees use different tools for different uses, suggesting that there may be cultural variations in tool use (Whiten et al. 1999). Does this ability imply some understanding of appropriate physical reasoning in these great apes and corvids?

Povinelli (2000) tested chimpanzees' understanding of how tools work, how using tools causes particular outcomes, and how certain objects are connected. He examined whether chimps understood that specific tools could only be used for specific jobs and that some tools were useless due to their physical properties. Surprisingly, his chimps were poor at almost all of the tasks they were presented, even when they were analogues of tool use in the wild. These experiments suggest that, although chimpanzees use tools, they may not understand the physical properties of the tools they are using. Povinelli (2000, 7) concluded that "chimpanzees do not represent abstract causal variables as explanations for why objects interact in the ways that they do."

These failures to demonstrate insight into the physical properties of tools are particularly intriguing in the light of some recent laboratory studies with the New Caledonian crows. When presented with a variety of sticks of different lengths and food positioned in a tube such that a stick was required in order to reach the food, the birds correctly chose the appropriate length of stick to push out the piece of food (Chappell and Kacelnik 2002). In a subsequent task, the crows were able to select the right diameter of tool (Chappell and Kacelnik 2004), suggesting that these birds have an advanced level of folk physics. Even more intriguingly, Weir and colleagues (2002) have shown that these tool-using crows can manipulate novel man-made objects to solve a problem. Two crows, Betty and Able, were presented with the problem of reaching food in a bucket that was only accessible by using a hook to pull the bucket up. Unfortunately, Able stole the bent wire and then dropped it somewhere out of Betty's reach. Betty found a piece of straight wire that was lying on the floor, bent this wire into a hook, and used it to lift up the bucket and reach the food!

Betty proceeded to do this successfully on nine out of the ten test trials.

Evidence of tool use and manufacture suggests that animals can sometime combine past experiences to produce novel solutions to problems. However, careful experimentation is required to establish whether the animal can flexibly exploit the tool in a way that suggests it can understand and reason about the causal relations between them. To date, there is no convincing evidence that animals do understand the physics of tools, but the most promising tool-using candidate, the New Caledonian crow, has yet to be tested. There is also recent evidence that one of the non-tool-using species of corvid, the rook, has some understanding of cause-and-effect relations in a modified tool task (Seed et al. 2006).

Conclusions and Implications

Much of the research on the evolution of mental characteristics has focused on the large-brained social primates because of their close evolutionary relationship to humans. The common assumption is that intelligence has evolved once within the primate lineage and that the complexities of social life led to an increase in mental abilities and to an expansion of the prefrontal cortex. However, intelligence may have evolved in other lineages of large-brained social animals such as cetaceans. In this chapter, we argue that corvids are large-brained social birds with mental abilities that are similar to great apes. As the last common ancestor to corvids and apes lived over 300 million years ago, we suggest that these similarities in intelligence must have developed through a process of convergence, rather than common ancestry (homology), as a result of adaptation to similar selective pressures. Furthermore, we suggest that this process of convergent evolution was driven by the requirement to solve comparable social and ecological problems.

The most recently evolved genera of corvids (*Corvus, Aphelocoma*) and apes (*Pan*) appeared at approximately the same geological time (five to ten million years ago). The late Miocene to Pliocene was a period of great environmental and climatic variability. This variability will have influenced food availability. As such, extant crows and apes may have had to adapt strategies for locating food dispersed in time and space, extracting food hidden

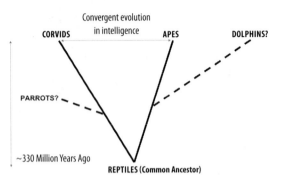

Figure 2. The proposed schema for the convergent evolution of intelligence in corvids and apes. The dotted lines denote the suggestion that this convergence in cognitive ability might also apply to other groups of birds and mammals, two obvious contenders being the parrots and dolphins respectively.

in cased substrates, and, thus, becoming innovative omnivorous generalist foragers. Such conditions will have had an effect on the organization of social groups. These ecological variables have been suggested to have played an important role in the evolution of ape cognition (Potts 2004), and we propose a similar scenario for the evolution of crow cognition. Interestingly, Lefebvre and colleagues found that flexibility in behavior, as measured by innovation rate, correlated with relative brain size in both birds and mammals. Furthermore, the corvids and apes displayed similar ratios of innovation rate to relative brain size, with members of the crow family having the highest values for birds and chimpanzees having the highest values for mammals (Lefebvre et al. 2004).

Marino (2002, 30) has argued that

Cetacean brains and primate brains represent alternative ways brains can increase in size and complexity and arrive at similar cognitive or even computational capacities. Therefore, this example implies that there may be general principles or "rules" that underlie the evolution of intelligence and that the specific way that a species arrives at a functional solution is not perhaps the only level at which to understand intelligence.

The case for corvids (and possibly also for parrots) is even more striking because the anatomical organization of the brain of birds and mammals is so different (Figure 2). Unlike the highly intelligent mammals, birds do not have a prefrontal cortex. We conclude that intelligence in both corvids and primates has evolved through a process of divergent brain evolution yet convergent mental evolution (see also Emery and Clayton 2004b). These findings have important implications for understanding the evolution of intelligence because they suggest that intelligence can evolve in the absence of a prefrontal cortex. Perhaps it only a matter of time until the galaxy of the corvids transcends the planet of the apes!

References

Barrett, L., P. Henzi, and R. Dunbar. 2003. Primate cognition: From "what now?" to "what if?" *Trends in Cognitive Sciences* 7: 494–97.

Beck, B. 1980. *Animal tool behavior: The use and manufacture of tools by animals.* New York: Garland.

Burish, M. J., H. Y. Kueh, and S. S.-H. Wang. 2004. Brain architecture and social complexity in modern and ancient birds. *Brain, Behaviour and Evolution* 63: 107–24.

Bugnyar, T., and B. Heinrich. 2005. Ravens, *Corvus corax*, differentiate between knowledgeable and ignorant conspecifics. *Proceedings of the Royal Society: B* 272: 1641–46.

Bugnyar, T., and K. Kotrschal. 2002. Observational learning and the raiding of food caches in ravens, *Corus corvax*: Is it "tactical" deception? *Animal Behaviour* 64: 185–95.

Byrne, R. W. 1997. The technical intelligence hypothesis: An additional evolutionary stimulus to intelligence? In *Machiavellian intelligence II: Extensions and evaluations*, ed. A. Whiten and R. W. Byrne, 289–311. Cambridge: Cambridge University Press.

Byrne, R. W., and A. Whiten. 1988. *Machiavellian intelligence: Social expertise and the evolution of intellect in monkeys, apes and humans.* Oxford, UK: Clarendon Press.

Chappell, J., and A. Kacelnik. 2002. Tool selectivity in a non-primate, the New Caledonian crow (*Corvus moneduloides*). *Animal Cognition* 5: 71–76.

———. 2004. Selection of tool diameter by New Caledonian crows *Corvus moneduloides*. *Animal Cognition* 7: 121–27.

Clayton, N. S., and A. Dickinson. 1998. Episodic-like memory during cache recovery by scrub jays. *Nature* 395: 272–78.

Clayton, N. S., T. J. Bussey, and A. Dickinson. 2003. Can animals recall the past and plan for the future? *Nature Reviews Neuroscience* 4: 685–91.

Clayton, N. S., J. M. Dally, J. D. J. Gilbert, and A. Dickinson. 2005. Food caching by western scrub-jays (*Aphelocoma californica*): A case of prospective cognition. *Journal of Experimental Psychology: Animal Behavioural Processes* 31: 115–24.

Clayton, N. S., and N. J. Emery 2007. The social life of corvids. *Current Biology* 17: R652-656.

Clayton, N. S., K. S. Yu, and A. Dickinson. 2003. Interacting cache memories: Evidence for flexible memory use by western scrub-jays (*Aphelocoma californica*). *Journal of Experimental Psychology: Animal Behavioural Processes* 29: 14–22.

Conway Morris, S. 2003. *Life's solution: Inevitable humans in a lonely universe.* Cambridge: Cambridge University Press.

Correia, S. P. C., D. M. Alexis, A. Dickinson, and N. S. Clayton. 2007. Western scrub-jays (*Aphelocoma californica*) anticipate future needs independently of their current motivational state. *Current Biology* 17: 856–61.

Dally, J. M., N. J. Emery, and N. S. Clayton. 2004. Cache protection strategies by western scrub-jays (*Aphelocoma californica*): Hiding food in the shade. *Proceedings of the Royal Society: B Biological Letters* 271: S387–90.

———. 2005. Cache protection strategies by western scrub-jays: Implications for social cognition. *Animal Behaviour* 70: 1251–63.

———. 2006. Food-caching western scrub-jays keep track of who was watching when. *Science* 312: 1662–65.

Darwin, C. 1872. *The expression of the emotions in man and animals*. Chicago: University of Chicago Press.

de Kort, S. R., S. P. C. Correia, D. M. Alexis, A. Dickinson, and N. S. Clayton. 2007. Food caching by western scrub-jays. *Journal of Experimental Psychology: Animal Behavior Processes*. 33, 361-70.

Dunbar, R. I. M. 1992. Neocortex size as a constraint on group size in primates. *Journal of Human Evolution* 20: 469–93.

———. 1998. The social brain hypothesis. *Evolutionary Anthropology* 6: 178–90.

Emery, N. J., and N. S. Clayton. 2001. Effects of experience and social context on prospective caching strategies by scrub jays. *Nature* 414: 443–46.

———. 2004a. Comparing the complex cognition of birds and primates. In *Comparative vertebrate cognition,* ed. L. J. Rogers and G. S. Kaplan, 3–55. The Hague: Kluwer Academic Press.

———. 2004b. The mentality of crows: Convergent evolution of intelligence in corvids and apes. *Science* 306: 1903–7.

———. 2005. Evolution of avian brain and intelligence. *Current Biology* 15: R1–5.

Emery, N. J., A. M. Seed, A. M. P. von Bayern, and N. S. Clayton. 2007. Cognitive adaptations of social bonding in birds. *Philosophical Transactions of the Royal Society of London B* 362: 489–505.

Hunt, G. R. 1996. Manufacture and use of hook-tools by New Caledonian crows. *Nature* 379: 249–51.

———. 2002. Tool use by the New Caledonian crow *Corvus moneduloides* to obtain Cerambycidae from dead wood. *Emu* 100: 109–14.

Hunt, G. R., and R. D. Gray. 2003. Diversification and cumulative evolution in New Caledonian crow tool manufacture. *Proceedings of the Royal Society London B* 270: 867–74.

Humphrey, N. K. 1976. The social function of intellect. In *Growing points in ethology*, ed. P. P. G. Bateson and R. A. Hinde, 303–17. Cambridge: Cambridge University Press.

Iwaniuk, A. N., and J. E. Nelson. 2003. Developmental differences are correlated with relative brain size in birds: A comparative analysis. *Canadian Journal of Zoology* 81: 1913–28.

Iwaniuk, A. N., K. M. Dean, and J. E. Nelson. 2005. Interspecific allometry of the brain and brain regions in parrots (Psittaciformes): Comparisons with other birds and primates. *Brain, Behavior and Evolution* 65: 40–59.

Jolly, A. 1966. Lemur social behaviour and primate intelligence. *Science* 153: 501–6.

Lefebvre, L., P. Whittle, E. Lascaris, and A. Finkelstein. 1997. Feeding innovations and forebrain size in birds. *Animal Behaviour* 53: 549–60.

Lefebvre, L., N. Nicolakakis, and D. Boire. 2002. Tools and brains in birds. *Behaviour* 139: 939–73.

Lefebvre, L., S. M. Reader, and D. Sol. 2004. Brains, innovations and evolution in birds and primates. *Brain, Behavior and Evolution* 63: 233–46.

Macintosh, N. J. 1988. Approaches to the study of animal intelligence. *British Journal of Psychology* 79: 509–25.

Marino, L. 2002. Convergence of complex cognitive abilities in cetaceans and primates. *Brain, Behavior and Evolution* 59: 21–32.

Milton, K. 1981. Distribution patterns of tropical plant foods as an evolutionary stimulus to primate mental development. *American Anthropologist* 83: 534–48.

Mogensen, J., and I. Divac. 1982. The prefrontal "cortex" in the pigeon: Behavioral evidence. *Brain, Behaviour and Evolution* 21: 60–66.

Parker, S. T., and K. R. Gibson. 1977. Object manipulation, tool use and sensorimotor intelligence as feeding adaptations in cebus monkeys and great apes. *Journal of Human Evolution* 6: 623–41.

Portmann, A. 1947. Etude sur la cérébralisation chez les oiseaux II: Les indices intra-cérébraux. *Alauda* 15: 1–15.

Potts, R. 2004. Paleoenvironmental basis of cognitive evolution in great apes. *American Journal of Primatology* 62: 209–28.

Povinelli, D. 2000. *Folk physics for apes: The chimpanzee's theory of how the world works.* Oxford: Oxford University Press.

Quammen, D. 1985. Has success spoiled the crow. In *Natural acts: A sidelong view of science and nature*, 30–31. New York: New York Press.

Raby, C. R., D. M. Alexis, A. Dickinson, and N. S. Clayton. 2007. Planning for the future by western scrub-jays. *Nature* 445: 919–21.

Rehkamper, G., H. D. Frahm, and K. Zilles. 1991. Quantitative development of brain and brain structures in birds (Galliformes and Passeriformes) compared to that in mammals (Insectivores and Primates). *Brain, Behavior and Evolution* 37: 125–43.

Reiner, A. 1986. Is prefrontal cortex found only in mammals? *Trends in Neuroscience* 9: 298–300.

Reiner, A., D. J. Perkel, L. L. Bruce, A. B. Butler, A. Csillag, W. Kuenzel, L. Medina, G. Paxinos, T. Shimizu, G. Striedter, M. Wild, G. F. Ball, S. Durand, O. Güntürkün, D. W. Lee, C. V. Mello, A. Powers, S. A. White, G. Hough, L. Kubikova, T. V. Smulders, K. Wada, J. Dugas-Ford, S. Husband, K. Yamamoto, J. Yu, C. Siang, and E. D. Jarvis. 2004. Revised nomenclature for avian telencephalon and some related brainstem nuclei. *Journal of Comparative Neurology* 473: 377–414.

Rilling, J. K., and T. R. Insel. 1999. The primate neocortex in comparative perspective using magnetic resonance imaging. *Journal of Human Evolution* 37: 191–223.

Sax, B. 2003. *Crow.* London: Reaktion Books Limited.

Seed, A. M. Tebbich, N. J. Emery, and N. S. Clayton. 2006. Investigating physical cognition in rooks (*Corvus frugilegus*). *Current Biology* 16: 697–701.

Semendeferi, K., A. Lu, N. Schenker, and H. Damasio. 2002. Humans and great apes share a large frontal cortex. *Nature Neuroscience* 5: 272–76.

Suddendorf, T., and M. C. Corballis. 1997. Mental time travel and the evolution of the human mind. *Genetic, Social, General Psychological Monographs* 123: 133–67.

———. 2007. The evolution of foresight: What is mental time travel and is it unique to humans? *Behavioral & Brain Sciences*.

Voronov, L. N., L. B. Bogoslovskaya, and E. G. Markova. 1994. A comparative study of the morphology of forebrain in corvidae in view of their trophic specialization. *Zoologicheskii Zhurnal* 73: 82–96.

Weir, A. A. S., J. Chappell, and A. Kacelnik. 2002. Shaping of hooks in New Caledonian crows. *Science* 297: 981.

Whiten, A., J. Goodall, W. C. McGrew, T. Nishida, V. Reynolds, Y. Sugiyama, C. E. G. Tutin, R. W. Wrangham, and C. Boesch, C. 1999. Cultures in chimpanzees. *Nature* 399: 682–85.

Wilson, B. J., N. J. Mackintosh, and R. A. Boakes. 1985. Transfer of relational rules in matching and oddity learning by pigeons and corvids. *Quarterly Journal of Experimental Psychology* 37B: 313–32.

8   SOCIAL AND CULTURAL EVOLUTION IN THE OCEAN

Convergences and Contrasts with Terrestrial Systems

Hal Whitehead

The Ocean and the Land

I live by the ocean. The windows on one side of the house look out on a complex landscape of rocks, grass, and trees, as well as products of human activity such as houses and roads. These, by and large, stay in place. Out the other side is the ocean, fluid and mobile. The ocean is mirror-like on some days, rough and wild on others—an environment that is quite predictable over meters and minutes but enormously variable over hundreds of kilometers and months. Steele (1985) notes that, in the ocean, environmental noise (after removing predictable cyclical variation: diurnal, lunar, annual) is largely "red" (greatest over large time and space scales), while, on land, it is more "white" (roughly constant over all scales, up to a century or continent or so). The fundamental contrasts between the two habitats are illustrated by the methods of the scientists who study them: terrestrial landscape ecologists plot habitats using geographical information systems, while their oceanographic counterparts use the partial differential equations of fluid dynamics to describe the marine environment.

The environment is the stage for evolution's play: organisms evolve to maximize their fitness given the environment. Two general environmental traits known greatly to influence evolution are connect-edness and variability (Grant 1986). In these respects, and many others, marine and terrestrial systems differ radically. And so, with such different sets, we might expect immense contrasts in both the action of the evolutionary play and its results, whether it occurs on land or in the ocean. And there are. Looking out my north-facing windows onto the land, most of the primary productivity I see is in large, long-lived, slowly reproducing spruce and maple trees. To the south, in the sea, the primary productivity is in microscopic diatoms and dinoflagellates. There are squid on one side, neutrally buoyant and forming large schools in an open three-dimensional habitat, and foxes on the other, negotiating trees and rocks and roads by themselves or in small groups, anchored by gravity. These are very different creatures in very different environments. Terrestrial and oceanic environments provide a tough challenge for convergence. When traits do converge, something remarkable has occurred.

Despite the radically different physical environments and forms of primary productivity on the land and in the ocean, as we move up the trophic web, convergences between oceanic and terrestrial animals begin to appear. The eyes of squid and foxes are an example (Conway Morris 2003). But at the trophic peaks and at the highest levels of biological organization, then the convergences between oceanic and terrestrial systems become particularly strong and provoking. A diatom and a spruce tree have little in common other than being autotrophs, but, as I will try to show, the social structures and cultures of sperm and killer whales have much in common with those of elephants and humans.

Phylogenetic constraints play a part in this. The diatom and maple are about as distantly related as any two organisms on Earth, whereas elephants, sperm whales, killer whales, and humans are all mammals and share all the constraints and advantages of the mammalian order, including backbones, air-breathing, warm blood, live birth, and lactation. But their common ancestor, perhaps one hundred million years ago, while possessing a backbone, air-breathing, warm-blooded with live birth and lactation, was small, likely socially and culturally primitive, certainly nothing like today's large and dominant mammals of land and ocean.

Social Convergence: Led by the Nose

Convergences between marine and terrestrial societies should be viewed from the perspective of social evolution. The social structure of a population of animals may be seen as the content, quality, and patterning of the relationships between its members, with relationships themselves representing the content, quality, and patterning of interactions between pairs (or possibly more) of individuals (Hinde 1976). The conventional wisdom of mammalian social evolution has it that relationships and interactions among females function to reduce predation on the females and their dependent young, to gain resources efficiently, and in intraspecific competition for territory and resources (Wrangham and Rubenstein 1986). Given the social structure of the females, males arrange their social behavior to assist in offspring care, if they are needed and can be fairly certain of paternity, and, then, to maximize mating potential (Clutton-Brock 1989). Thus, we expect that different social systems should have resulted from the very different patterns of resource distribution in terrestrial and marine environments. For instance, food patches for open-ocean mammals tend to be large, ephemeral, and not defensible, while, for land mammals, they are smaller and more predictable and can often be defended economically.

There are contrasts between mammalian social structures on land and in the ocean. Whereas territoriality is common among terrestrial mammals, it is absent, as far as we know, among mammals of the three-dimensional, and much less defensible, ocean (Whitehead 2003b). Marine mammals possess one social system yet to be found on land: among "resident" (fish-eating) killer whales, pilot whales, and maybe other cetacean species, both males and females spend their lives grouped with their mother and her relatives (Connor et al. 1998). There is no dispersal. We think that this works in the ocean because, with no territoriality and easy movement, a male will incidentally meet plenty of unrelated females with whom he can potentially mate while staying with his mother and gaining all the advantages of living in a close-knit family group. On land, such social benefits cannot compensate for the substantial mating opportunities lost by being tied to mother.

But there are also convergences between the social systems of marine and terrestrial mammals. Connor et al. (2000) discuss apparent convergences in mating strategies between dolphins and apes. For instance, females of both bottlenose dolphins (*Tursiops spp.*) and chimpanzees (*Pan troglodytes*) live in fission-fusion societies, whereas males form alliances with other males to guard receptive females temporarily.

However, the most comprehensive convergence between marine and terrestrial mammals is between the species that I study, the sperm whale (*Physeter macrocephalus*), and the elephants (*Elephas maximus* and *Loxodonta spp.*) (Weilgart, Whitehead, and Payne 1996; Whitehead 2003b). Termed "the Colossal Convergence" by an editor at *American Scientist*, it features a wide range of traits in which elephants are more similar to sperm whales than they are to other terrestrial mammals and sperm whales are more similar to elephants than other marine mammals. In both species, females live in largely matrilineal social units of about eleven animals within which there is communal care for the young and communal defense against predators. These social units aggregate to form larger social structures, including groups of about twenty animals. Males leave their mothers' social units, segregate from the females, and grow to become much larger than their mothers. In their late twenties, the males return to the habitat of the females to mate, roving between the female units, competing with each other and being selected by the females.

There are additional, nonsocial, parallels between sperm whales and elephants. For instance, the species have very similar life histories and are nonterritorial and quite mobile. And both are extreme in other respects. These include body size and brain size: the sperm whale has the largest brain of all species, and the elephant the largest among land animals. Another parallel is in ecological success. Elephants, due to their size, numbers, and feeding methods can restructure habitats (Laws 1970), while the world's sperm whale population, even though substantially reduced by whaling, currently removes about as much biomass from the oceans as all human marine fisheries combined (Whitehead 2003b).

This convergence between a terrestrial herbivore and a marine teuthivore (the sperm whale principally eats deep-water squid) is striking, especially given the radical contrasts in their habitat and food: quite a puzzle for the evolutionary biologist. I have proposed that the key convergence is in the nose (Whitehead 2003b). The elephant's trunk is a most distinctive, unusual, and useful organ. And the sperm whale has a similarly impressive nose: the spermaceti organ is a massive, complex, oil-filled sack. It is situated above the upper jaw, and around it loop the nasal passages, one of which terminates in the blowhole through which the whale breathes. The spermaceti organ spans about 25 percent of the sperm whale's body and is sheathed in powerful muscle. Recent research (especially that of Møhl et al. 2000) has shown that this organ forms the most powerful sonar in the natural world. Like the elephant's trunk, it gives its bearer an enormously valuable tool in the struggle to wrest nutrition from the environment and an advantage over competing species.

My evolutionary scenario starts with primitive, and smaller, elephants and sperm whales developing the trunk and the spermaceti organ, respectively. With their wonderful noses, elephants and sperm whales became ecologically dominant. Similar species disappeared as they were outcompeted—there is now nothing much like either an elephant or sperm whale. Then, with intraspecific competition regulating individual fitness, life history processes slowed to produce the pattern that used to be called "K-selection," before that term fell from favor (Stearns 1992). Slow life histories and sociality feed back upon one another (Horn and Rubenstein 1984), as long lives promote social bonds and as social bonds reduce mortality through cooperative defense of females and their vulnerable offspring, communal environmental knowledge, and other mechanisms. So, we have ecologically successful, long-lived, social species. These include humans, chimpanzees, bottlenose dolphins, and some social birds.

But in the cases of the sperm whale and the elephant, this positive feedback process was, in some respects, wound further: the powerful feeding apparatuses contained in the noses of these species permit high rates of energy gain and very large body size. Large animals are safer, so tend to live longer, but can also more easily afford large brains and complex cognition, a feature of advanced sociality (Byrne and Whiten 1988). And so, according to my scenario, we have the principal elements of the elephant–sperm whale colossal convergence.

A possible implication of this model is that a species that develops a novel and functionally advantageous generalized foraging mechanism might follow the colossal convergence. However, there may be certain prerequisites. The forced mother–infant dependency during lactation links sociality and life history processes more tightly for mammals than for species with more independent offspring. An initial largish size could also be significant, as a significant feeding innovation in very small animals might not be advantageous when scaled up. It may also be significant that both elephants and sperm whales, although very large and powerful, have potentially dangerous predators, especially on the young: lions and killer whales, respectively. The presence of such predators likely favored sociality, as well as large size. Given these restrictions, are there any other fairly large mammals that have developed unusual advantageous generalized feeding mechanisms? I cannot think of any, so perhaps the colossal convergence is restricted to these two.

Cultural Convergence: Ultrasociality

Evolution through natural selection requires a transmission mechanism, so that phenotypes are passed between individuals. Almost always, evolutionary biologists are concerned with genetic transmission, but there are other possibilities, of which culture is the most significant (Maynard Smith 1989), particularly for humans (Richerson and Boyd 2004). Culture can be defined in many ways, but, from an evolutionary perspective, the key elements are that individuals develop behavior patterns or gain information that affects behavior from one another through social learning. So, I define culture as behavior or information shared by members of a population or subpopulation and transmitted by some form of social learning (Rendell and Whitehead 2001a). As a transmission mechanism, culture has some similarities with genes. Cultural phenotypes can mutate and evolve and are subject to the natural selection of both cultural variants and culture-bearers; culture often leads to adaptive behavior. However, there are some important differences (see Boyd and Richerson 1985): individuals can receive culture from a range of donors in addition to their parents; they can choose which culture to adopt; and their own experiences and behavior can influence the form of culture that is transmitted to other individuals, so acquired characters can be inherited. Thus, cultural evolution has characteristics not found in genetic evolution. One of these is group selection, that behavior evolves for the good of the group: hard to achieve through genes but quite simple culturally (Boyd and Richerson 1985).

Much of how humans behave, almost all of what we produce, and, some would say, most of what we are results culture. Many believe that culturally we are unique or at least highly extreme. No other animals have produced anything like our books, aircraft, or music. But human "hyper-culture" is quite recent. For instance, human tools were few and stere-otyped until about half a million years ago, but then began to diversify, increasingly rapidly in recent centuries (Richerson and Boyd 2004). In the ocean, there is little or no tool use (Rendell and Whitehead 2001a), but we seem to find cultural convergence in a quite different area.

In addition to being hypercultural, modern humans are ultrasocial. In our nation-states, ethnic groups, religions, armies, and other large-scale groupings, we "organize cooperation on a far larger scale than our primate relatives" (Richerson and Boyd 1998, 92). The members of the nation-state of China (1.3 billion humans) cooperate to form a society that is in com-petition, in some respects, with other nation-states. In contrast to other cases of large-scale cooperation among animals, such as the colonial inver-tebrates and social insects, the individual members of large-scale coopera-tive groups in humans are not generally closely genetically related. Human societies have evolved well beyond the limits (a few tens of individuals) set by the mechanisms that are thought to structure cooperation in other ani-mal societies with diploid genetic systems: kinship and reciprocity (Rich-erson and Boyd 2004). So, from the perspective of a zoologist studying animal societies, the evolution of human ultrasociality appears both an exception and a puzzle.

Richerson and Boyd (1998) have produced a convincing scenario for the evolution and existence of this human ultrasociality. They believe that humans evolved sophisticated cultural capacity (social learning) that was adaptive in the highly variable Pleistocene environments. One form of cultural transmission is to adopt the commonest form of behavior present in the population, to conform. This will frequently be adaptive to an indi-vidual, especially one that lives in a cooperative society within a varia-ble environment (Boyd and Richerson 1985). Conformism can lead to the structuring of a population into culturally marked groups, with which individuals associate themselves (Boyd and Richerson 1985). This can result in the structuring of populations on very large scales, with little or no genetic basis, as in human nation-states.

Richerson and Boyd (1998) believed the cultural form of ultrasociality has only arisen once on Earth, in humans. I am not so sure. The groups of female sperm whales that we study communicate using Morse code–like patterns of clicks, called "codas" (Watkins and Schevill 1977). Different groups may have different coda repertoires (Weilgart and Whitehead 1997). Off the Galapagos Islands, from where we have most data, we mainly hear two principal patterns of coda: "regular" codas such as "click-click-click-click-click," and "plus-one" codas like "click-click-click-[pause]-click." Some social units principally make regular codas, some plus-one codas, and no unit has changed repertoire over the years we have studied them, so we classify them into the "regular" clan and "plus-one" clan (Rendell and Whitehead 2003). Although units of both clans are present off the Galapagos, each unit seems to group only with other units of its own clan.

So what produces these differences in vocal repertoire? The clans have very considerable genetic overlap, especially in nuclear markers, strongly indicating that the cause is not genetic (Rendell and Whitehead 2003; Whitehead 2003b). Each clan contains the whole range of ages and sex classes of sperm whales, except mature and maturing males (who disassociate from the females at about age six), so clans do not map onto ontogenetic or gender differences. And both clans are found in the same areas, so environmental differences plus individual learning cannot be the cause. This process of eliminating genetic, ontogenetic, and environmental causes leaves culture—social learning—as the only likely cause of the clan-specific coda repertoires (Rendell and Whitehead 2003; Whitehead 2003b).

But there is more to each clan than just a distinctive pattern of clicks and within-clan attraction. The two clans differ in movement patterns and habitat use: the "plus-one" units have substantially straighter paths, so that, despite similar speeds of swimming through the water, they have moved about twice as far over twelve-hour periods than units of the "regular" clan; and the "plus-one" units are found, on average, 20 kilometers further from the islands than the "regular" units (Whitehead and Rendell 2004). The clans also differ in feeding success (which we measure by recording the rates of defecation): under normal Galapagos conditions, units of the "regular" generally do substantially better; but when El Nino strikes, the waters warm, and conditions become very unfavorable for most marine life, including sperm whales, then the "plus-one" units have higher feeding success (Whitehead and Rendell 2004).

We have studied sperm whales across much of the South Pacific and have found social units of both the "regular" and "plus-one" clans across wide areas, as well as three to four other clans (Rendell and Whitehead 2003). Each clan has a range that spans about 2,000–10,000 kilometers and shares that range with other clans (Rendell and Whitehead 2003). From what we know of sperm whale population sizes, we can estimate that each clan has roughly tens of thousands of members. So sperm whale populations are structured culturally on a huge scale.

The similarities between human ethnic groups and sperm whale clans include distinctive vocal repertoires; distinctive behavior, including foraging methods; preference for associating with members of ones own ethnic group or clan in multicultural situations; large spatial scales; and numbers in at least the thousands. Neither kin selection nor reciprocity seems sufficient mechanisms to have produced these patterns. I suspect that cultural conformism and cultural group selection have been at work in producing sperm whale ultrasociality, as with humans (Richerson and Boyd 1998).

Are there other species where culture might have driven large-scale sociality? I suspect not on land, where territoriality and movement restrictions will have generally limited the possibilities for very large-scale sociocultural structures. However, the relatively mobile (Thouless 1995) and likely culturally advanced (McComb et al. 2001) elephant is perhaps the best candidate, especially given its other similarities with sperm whales (see above).

In the ocean, there is good evidence for large-scale, and likely culturally driven, social structures in killer whales. Killer whale populations are structured into a number of sociocultural levels. One of these is the clan (Ford 1991). Killer whale clans, like those of sperm whales (the name "clan" was taken from killer whale terminology when these structures were discovered in sperm whales; Rendell and Whitehead 2003), consist of a number of matrilineally based social units that have a common vocal dialect but share habitat with units from other clans (Ford 1991). However, there are differences between clans in sperm and killer whales: clan-specific, nonvocal attributes have not yet been described in killer whales, and killer whale clans are smaller, spanning about 1,000 kilometers and containing about one hundred members (Yurk et al. 2002), as compared with the tens of thousands of kilometers and tens of thousands of members suggested for sperm whales (Rendell and Whitehead 2003). So perhaps killer whale clans do not constitute "ultrasociality." However, there are larger-scale levels to killer whale society: "communities," containing a few hundred animals that are distinctive both vocally and in some behavioral respects: for instance, the "southern resident community" has a distinctive "greeting ceremony" when groups meet (the two groups line up facing one another for ten to thirty seconds before approaching and mingling [Osborne 1986]). There are small but consistent genetic differences between communities (Barrett-Lennard 2000), and ranges of different communities overlap little, so genetic and environmental causes of behavioral differences are not as easily dismissed as with killer and sperm whale clans. At the highest level and largest scales, there are sympatric "types" of killer whales (including fish-eating "residents" and mammal-eating "transients" in the eastern North Pacific) that differ in many attributes, including diet, social system, vocalizations, morphology, and genetics (Baird 2000). These types are large-scale and profound divisions of killer whales (Baird 2000), but the differences between types are so substantial that they may constitute subspecies, and so can hardly be considered sociocultural entities, even though cultural differences are suspected to have been fundamental in their evolution (Boran and Heimlich 1999, Baird 2000).

There are other species of largish toothed whales that seem to have matrilineally based social systems and other similarities with killer and sperm whales, such as pilot whales (*Globicephala* spp.). Although these have been studied little, it is not unreasonable to predict that they too might have culturally structured populations (Rendell and Whitehead 2001b) and perhaps ultrasociality.

Thus, I conclude this section by suggesting that, in some fundamental ways (including causal mechanisms), human ultrasociality is not unique. Moreover, since humans became ultrasocial over only the last few tens of thousands of years (Richerson and Boyd 1998), whereas sperm whales have changed little over a few million years (Rice 1989), it is unlikely that we were the first ultrasocial species.

Cultural Hitchhiking: Gene-culture Coevolution

The overwhelming majority of the convergences discussed by Conway Morris (2003) are believed to be genetically driven. In the previous section, I have suggested culturally mediated convergences. Here I will link the two in a counterintuitive manner: I believe that convergences in genetic attributes of some mammals of both the terrestrial and oceanic environments may have a common cultural mechanism.

Populations of widespread and numerous organisms are expected to possess relatively high genetic diversity. When they do not, something unusual has happened. As molecular geneticists began to publish the results of surveys of the diversity of whale genes in the mid-1990s, there were surprises. One of the most dramatic was that the diversity of mitochondrial DNA diversity, which is inherited maternally, was many-fold lower in sperm whales, killer whales, and the two pilot whale species than in other whale and dolphin species with similar population sizes and geographic ranges (Whitehead 1998, 2003a). These were the only four cetacean species in the twenty-species sample known to have a matrilineal social system, in the sense that females spend most of their lives grouped with their mothers during their shared lifetimes. The low diversity of mitochondrial DNA in these species has attracted a range of explanations. The standard causes for lower genetic diversity—population bottlenecks and selection—do not work well in this case: why should just the matrilineal whales be subject to bottlenecks or selection? More convincing are scenarios that treat the matrilineal groups as the units of genetic change in the population analysis, thus lowering the effective population size and expected genetic diversity (Siemann 1994; Amos 1999; Tiedemann and Milinkovitch 1999). However, for these explanations to work, the animals within a matrilineal group need to breed and/or die together, and there must be much environmental variation; as a result, the whole population is likely to go extinct (Whitehead 2005).

I have suggested that culture may be responsible for such unusually low genetic diversities through the mechanism of cultural hitchhiking (Whitehead 1998, 2005). If populations are culturally divided into sets, such as sperm whale "clans," with distinctive, selectively important cultures, then the process of cultural group selection can reduce diversity in those neutral genes that are transmitted in parallel to the cultural traits. To visualize this, imagine an ocean populated by matrilineally based clans of whales, so that a female almost always stays in the same clan as her mother. In such a situation, the different clans will tend to have different distributions of mitochondrial DNA haplotypes (as seems to be the case for the sperm whale clans; Rendell and Whitehead 2003), and each will possess less diversity than the entire population. The clans compete against one another for the resources of the ocean. If a good idea emerges and spreads in one clan, then its members will have better-than-average fitness so that the clan gradually takes over much of the ocean. As it does so, the genetic diversity of the whale population will fall towards that of the "good-idea" clan. I have shown, using computer models, that this scenario for reduced genetic diversity is expected under quite a range of clan sizes, rates of genetic mutation, cultural innovation, interclan migration, and interclan cultural assimilation (Whitehead 2005).

Another of the species discussed earlier in this chapter has remarkably low levels of genetic diversity. Like sperm, killer, and pilot whales, humans are widespread and numerous. They also have remarkably low levels of genetic diversity at some loci. Some chimpanzee groups have greater mitochondrial DNA diversity than the entire human species (Gagneux et al. 1999), and human Y-chromosome DNA, inherited paternally, is even more unexpectedly depauperate (Thomson et al. 2000). These low levels of genetic diversity seem to date from some process in the late Pleistocene. As with the whales, explanations using bottlenecks or selection are not very satisfactory (Harding 1999). So, my colleagues and I modified the cultural hitchhiking model of the whales to refer to a population of late Pleistocene hunter-gatherers: mobile matrilineal clans were replaced by largely patrilineal and territorial tribes (Whitehead, Richerson, and Boyd 2002). Once again, the cultural evolution was able to decimate genetic diversity, although the human model needed a higher rate of cultural innovation (Whitehead, Richerson, and Boyd 2002).

That cultural hitchhiking works in computer models with realistic parameters for both matrilineal whales and patrilineal humans is interesting but does not prove that the process has been part of the evolutionary history of any species. We need tests, but, unfortunately, no one has yet come up with a test, or tests, that can convincingly separate cultural hitchhiking from all competing hypotheses of selection, bottlenecks, and group-specific demographic effects (Whitehead 2005). With the bias of an originator, I think that cultural hitchhiking is the most plausible explanation for low genetic diversities in both humans and the matrilineal cetaceans. If so, this is another case of high-level convergence between marine and terrestrial systems and a counterargument to the case that gene-culture coevolution is restricted to humans (Feldman and Laland 1996).

Convergence between Land and Ocean

I have described a few of the convergences between the cetaceans and large, social, terrestrial mammals, especially the anthropoid apes and elephants. There are others (see Conway Morris 2003), including:

- Levels of encephalization: primates and cetaceans are the most encephalized of mammal orders; following humans, some cetacean species are the next most encephalized of mammal species (Marino 1998).
- On some cognitive tests, such as the mirror-mark test and the ability to use syntax, some cetaceans and some primates (especially the anthropoid apes) perform similarly and better than all other animals so far tested (Herman, Pack, and Wood 1994; Reiss and Marino 2001).
- Dolphins, like humans, may generalize rules and develop abstract concepts, such as the concepts of novelty and imitation (Pryor, Haag, and O'Reilly 1969; Herman, Pack, and Wood 1994; Herman and Pack 2001).
- Menopause, when reproductive function terminates well before the expected age of death (Whitehead and Mann 2000), is so far only known from killer whales (Olesiuk, Bigg, and Ellis 1990), pilot whales (Marsh and Kasuya 1986) and humans, although it may well be present in other cetacean species, especially those with matrilineal social systems, including sperm whales (Marsh and Kasuya 1986; Whitehead 2003b). As such, it seems to correlate quite closely with some of the social, cultural, and cognitive convergences discussed above. It is probable that, in these species, females gain greater

inclusive fitness by ceasing reproduction, rather than attempting to produce surviving offspring right up until their deaths. Clearly, a reduction in the probability of producing surviving offspring with age plays a part in this, but there must also be some fitness benefit to forgoing reproduction altogether. Current theories suggest that menopause may be beneficial by increasing the survival of a female's current offspring, or other relatives, especially grandchildren, through increased care and/or because extending a female's life by removing the costs and dangers of reproduction increases the access of her relatives to important cultural knowledge that she may have assimilated during her long life (Whitehead and Mann 2000).

Cognitive and psychological convergences between cetaceans and primates are significant in their own right. But psychological abilities, such as "theory of mind," may be prerequisites for the development of conformist cultures (Dunbar 2001; Richerson and Boyd 2004), which seem to have driven human, and sperm whale, ultrasociality, as well as perhaps cultural hitchhiking.

This is all true convergence. We can be confident that the small mammal that was the common ancestor of cetaceans, apes, and elephants all those millions of years ago did not possess anything like the social, cultural, or cognitive complexity, the life-history characteristics, or ecological success of recent humans, apes, elephants, or cetaceans. Despite the radically different physical structuring of the ocean and the land, despite the contrasts in the forms of the primary producers, despite their difference in dimension and the potential mobility of the inhabitants, we have remarkable convergence in the social, cognitive, and cultural lives of the mammals at the trophic apices.

No other order has been able to make the same kind of impact in both habitats. There are seabirds as well as land birds, but both breed on land; there are terrestrial and marine amphibians, but, in both habitats, they are ecologically overshadowed by the mammals. Crustaceans and fishes are important in aquatic habitats and insects in terrestrial ones, but none makes much of an impact when it crosses the boundary.

So I think the ocean-land convergences may be rooted in some mammalian characteristics such as high metabolic rates and the excellent parental care allowed by internal gestation and lactation. For instance, parental care and alloparental care are seen by some as being fundamental to complex social structure (Emlen 1991) and prolonged adult-infant dependency also sets the stage for cultural transmission of behavior (Richerson and Boyd 2004). Additionally, high metabolic rates are probably necessary for advanced brains.

So, mammals have the prerequisites for evolving a set of interdependent social, cognitive, psychological, and cultural capacities, those common to the great apes, elephants, and some odontocete cetaceans, as well the ecological and life-history characteristics that often go along with them. Not all mammals have taken this route (e.g., prosimians), and some have traveled part way (e.g., canids, phocids). There are also nonmammalian species, especially birds, that have social, cognitive, cultural, and life-history characteristics in common with the apes, odontocetes and elephants (see Nicola Clayton and Nathan Emery's chapter in this collection).

In summary, this comparison of the radically differently structured worlds of land and ocean indicates, very strongly I think, that, as we move up the trophic levels and levels of biological structure, evolution becomes channeled into particular paths, in which social relationships, cognition, and culture interact and may feed back into ecological, life history, and even genetic convergences. This, in turn, suggests a program of research (abhorred by some of my marine mammalogist colleagues; Tyack 2001, Boyd 2004), namely, of using characteristics of socially and cognitively advanced terrestrial vertebrates, such as elephants, chimpanzees, and humans, as working hypotheses in our explorations of life in the oceans. Do whales and dolphins have language? Current evidence suggests not, but there is much we do not know (Tyack 1999). Are they creative for creativity's sake? Do they have some sense of religion?

Acknowledgments

Many thanks to Simon Conway Morris and the Templeton Foundation for inviting me to the Castel Gandolfo symposium, to the Vatican Observatory, and especially to Father George Coyne for hosting it, and to Simon Conway Morris and Linda Weilgart for constructive comments on drafts.

References

Amos, W. 1999. Culture and genetic evolution in whales. *Science* 284: 2055a.

Baird, R. W. 2000. The killer whale—foraging specializations and group hunting. In *Cetacean societies*, ed. J. Mann, R. C. Connor, P. Tyack, and H. Whitehead, 127–53. Chicago: University of Chicago Press.

Barrett-Lennard, L. 2000. Population structure and mating patterns of killer whales (*Orcinus orca*) as revealed by DNA analysis. PhD diss., University of British Columbia, Vancouver, Canada.

Boran, J. R., and S. L. Heimlich. 1999. Social learning in cetaceans: Hunting, hearing and hierarchies. *Symposia of the Zoological Society, London* 73: 282–307.

Boyd, I. L. 2004. Culture among sperm whales? *Science* 302: 990.

Boyd, R., and P. Richerson. 1985. *Culture and the evolutionary process.* Chicago: University of Chicago Press.

Byrne, R., and A. Whiten, eds. 1988. *Machiavellian intelligence.* Oxford: Clarendon.

Clutton-Brock, T. H. 1989. Mammalian mating systems. *Proceedings of the Royal Society of London B* 236: 339–72.

Connor, R. C., A. J. Read, and R. Wrangham. 2000. Male reproductive strategies and social bonds. In *Cetacean societies*, ed. J. Mann, R. C. Connor, P. L. Tyack, and H. Whitehead, 247–69. Chicago: University of Chicago Press.

Connor, R. C., J. Mann, P. L. Tyack, and H. Whitehead. 1998. Social evolution in toothed whales. *Trends in Ecology and Evolution* 13: 228–32.

Conway Morris, S. 2003. *Life's solution. Inevitable humans in a lonely universe.* Cambridge: Cambridge University Press.

Dunbar, R. I. M. 2001. Do how do they do it? *Behavioral and Brain Sciences* 24: 332–33.

Emlen, S. T. 1991. Evolution of cooperative breeding in birds and mammals. In *Behavioural ecology. An evolutionary approach*, 3rd ed., ed. J. R. Krebs and N. B. Davies, 301–37. Oxford: Blackwell.

Feldman, M. W., and K. N. Laland. 1996. Gene-culture coevolutionary theory. *Trends in Ecology and Evolution* 11: 453–57.

Ford, J. K. B. 1991. Vocal traditions among resident killer whales (*Orcinus orca*) in coastal waters of British Columbia. *Canadian Journal of Zoology* 69: 1454–83.

Gagneux, P., C. Wills, U. Gerloff, D. Tautz, P. A. Morin, C. Boesch, B. Fruth, G. Hohmann, O. A. Ryder, and D. S. Woodruff. 1999. Mitochondrial sequences show diverse evolutionary histories of African hominids. *Proceedings of the National Academy of Sciences USA* 96: 5077–82.

Grant, P. R. 1986. Ecology and evolution of Darwin's finches. Princeton, NJ: Princeton University Press.

Harding, R. M. 1999. More on the X files. *Proceedings of the National Academy of Sciences USA* 96: 2582–84.

Herman, L. M., and A. A. Pack. 2001. Laboratory evidence for cultural transmission mechanisms. *Behavioral and Brain Sciences* 24: 335–36.

Herman, L. M., A. A. Pack, and A. M. Wood. 1994. Bottlenose dolphins can generalize rules and develop abstract concepts. *Marine Mammal Science* 10: 70–80.

Hinde, R. A. 1976. Interactions, relationships and social structure. *Man* 11: 1–17.

Horn, H. S., and D. I. Rubenstein. 1984. Behavioural adaptations and life history. In *Behavioural ecology. An evolutionary approach*, 2nd ed., ed. J. R. Krebs and N. B. Davies, 279–98. Oxford: Blackwell Science Publications.

Laws, R. M. 1970. Elephants as agents of habitat and landscape change in East Africa. *Oikos* 21: 1–15.

Marino, L. 1998. A comparison of encephalization between odontocete cetaceans and anthropoid primates. *Brain, Behaviour and Evolution* 51: 230–38.

Marsh, H., and T. Kasuya. 1986. Evidence for reproductive senescence in female cetaceans. *Reports of the International Whaling Commission* (Special Issue) 8: 57–74.

Maynard Smith, J. 1989. *Evolutionary genetics*. Oxford: Oxford University Press.

McComb, K., C. Moss, S. M. Durant, L. Baker, and S. Sayialel. 2001. Matriarchs as repositories of social knowledge in African elephants. *Science* 292: 491–94.

Møhl, B., M. Wahlberg, P. T. Madsen, L. A. Miller, and A. Surlykke. 2000. Sperm whale clicks: Directionality and source level revisited. *Journal of the Acoustical Society of America* 107: 638–48.

Olesiuk, P., M. A. Bigg, and G. M. Ellis. 1990. Life history and population dynamics of resident killer whales (*Orcinus orca*) in the coastal waters of British Columbia and Washington State. *Reports of the International Whaling Commission* (Special Issue) 12: 209–43.

Osborne, R. W. 1986. A behavioral budget of Puget Sound killer whales. In *Behavioral biology of killer whales*, ed. B. Kirkewold and J. S. Lockard, 211–49. New York: A. R. Liss.

Pryor, K. W., R. Haag, and J. O'Reilly. 1969. The creative porpoise: Training for novel behavior. *Journal of the Experimental Analysis of Behavior* 12: 653–61.

Reiss, D., and L. Marino. 2001. Mirror self-recognition in the bottlenose dolphin: A case of cognitive convergence. *Proceedings of the National Academy of Sciences* USA 98: 5937–42.

Rendell, L., and H. Whitehead. 2001a. Cetacean culture: Still afloat after the first naval engagement of the culture wars. *Behavioral and Brain Sciences* 24: 360–73.

———. 2001b. Culture in whales and dolphins. *Behavioral and Brain Sciences* 24: 309–24.

———. 2003. Vocal clans in sperm whales (*Physeter macrocephalus*). *Proceedings of the Royal Society of London B* 270: 225–31.

Rice, D. W. 1989. Sperm whale. *Physeter macrocephalus Linnaeus*, 1758. In *Handbook of marine mammals*, ed. S. H. Ridgway and R. Harrison, 4:177–233. London: Academic Press.

Richerson, P. J., and R. Boyd. 1998. The evolution of human ultrasociality. In *Indoctrinability, ideology and warfare*, ed. I. Eibl-Eibesfeldt and F. K. Salter, 71–95. London: Berghahn Books.

———. 2004. *Not by genes alone: How culture transformed human evolution*. Chicago: University of Chicago Press.

Siemann, L. A. 1994. Mitochondrial DNA sequence variation in North Atlantic longfinned pilot whales, *Globicephala melas*. PhD diss., Massachusetts Institute of Technology, Cambridge, Massachusetts.

Stearns, S. C. 1992. *The evolution of life histories*. Oxford: Oxford University Press.

Steele, J. H. 1985. A comparison of terrestrial and marine ecological systems. *Nature* 313: 355–58.

Thomson, R., J. K. Pritchard, P. Shen, P. J. Oefner, and M. W. Feldman. 2000. Recent common ancestry of human Y chromosomes: Evidence from DNA sequence data. *Proceedings of the National Academy of Sciences* USA 97: 7360–65.

Thouless, C. R. 1995. Long distance movements of elephants in northern Kenya. *African Journal of Ecology* 33 :321–34.

Tiedemann, R., and M. Milinkovitch. 1999. Culture and genetic evolution in whales. *Science* 284: 2055a.

Tyack, P. 1999. Communication and cognition. In *Biology of marine mammals*, ed. J. E. Reynolds and S. A. Rommel, 287–323. Washington, DC: Smithsonian Institution Press.

Tyack, P. L. 2001. Cetacean culture: Humans of the sea? *Behavioral and Brain Sciences* 24: 358–59.

Watkins, W. A., and W. E. Schevill. 1977. Sperm whale codas. *Journal of the Acoustical Society of America* 62: 1486–90.

Weilgart, L., and H. Whitehead. 1997. Group-specific dialects and geographical variation in coda repertoire in South Pacific sperm whales. *Behavioural Ecology and Sociobiology* 40: 277–85.

Weilgart, L., H. Whitehead, and K. Payne. 1996. A colossal convergence. *American Scientist* 84: 278–87.

Whitehead, H. 1998. Cultural selection and genetic diversity in matrilineal whales. *Science* 282: 1708–11.

———. 2003a. Society and culture in the deep and open ocean: The sperm whale. In *Animal social complexity: Intelligence, culture and individualized societies*, ed. F. B. M. de Waal and P. L. Tyack, 444–64. Cambridge, MA: Harvard University Press.

———. 2003b. *Sperm whales: Social evolution in the ocean*. Chicago: University of Chicago Press.

———. 2005. Genetic diversity in the matrilineal whales: Models of cultural hitchhiking and group-specific non-heritable demographic variation. *Marine Mammal Science* 21: 58–79.

Whitehead, H., and J. Mann. 2000. Female reproductive strategies of cetaceans. In *Cetacean societies*, ed. J. Mann, R. Connor, P. L. Tyack, and H. Whitehead, 219–46. Chicago: University of Chicago Press.

Whitehead, H., and L. Rendell. 2004. Movements, habitat use and feeding success of cultural clans of South Pacific sperm whales. *Journal of Animal Ecology* 73: 190–96.

Whitehead, H., P. J. Richerson, and R. Boyd. 2002. Cultural selection and genetic diversity in humans. *Selection* 3: 115–25.

Wrangham, R. W., and D. I. Rubenstein, eds. 1986. *Ecological aspects of social evolution*. Princeton, NJ: Princeton University Press.

Yurk, H., L. Barrett-Lennard, J. K. B. Ford, and C. O. Matkin. 2002. Cultural transmission within maternal lineages: Vocal clans in resident killer whales in southern Alaska. *Animal Behaviour* 63: 1103–19.

9 THE ILLUSION OF PURPOSE IN EVOLUTION

A Human Evolutionary Perspective

Robert A. Foley

Introduction

There are a number of concepts that are no longer acceptable parts of the evolutionary biologist's intellectual armory. The themes of this book and the symposium on which it is based read like a library of books that would be on the list of any evolutionary biologist's equivalent of the *Index Librorum Prohibitorum*—progress, purpose, direction, end point, design, plan, inevitability, increased complexity. These are all concepts that, by and large, are no longer part of the working concepts of evolutionary biology. There are good reasons for this. The acceptance of the modern synthesis of evolutionary theory swept aside many of the barnacles that had accrued to Darwinian thought in its first hundred years, replacing them with a much more straightforward and mechanistic set of processes. Evolutionary change, under this model, was the product of selection, with no directionality other than that determined by the selective environment. Design brought about by either external "powers" or evolvability could play no part. Adaptations—the key element in modern Darwinian theory—were by definition relative; therefore, to rank these in any absolute sense that could be interpreted in terms of progress was no longer tenable. These simplifications and clarifications gave rise to the power—and the beauty in terms of its simplicity—of modern evolutionary biology, summed up eloquently in Richard Dawkins' term the blind watchmaker (Dawkins 1987). Furthermore, while the modern synthesis removed all these concepts through simplification, the macro-evolutionary debate of the late twentieth century castigated many of them even further, although less through simplification than through placing greater and greater emphasis on the role of chance and on the inertial effect of history—"contingency" in the terms of Gould and others (Gould 1977).

Any reasoned look at the history of evolutionary debate, then, would make the idea of restoring a purpose to evolution one of the more daunting scientific challenges. Indeed, I can start by stating that it is doomed as an enterprise if the intention—purpose, perhaps—is to bring the idea of an ultimate point, end point, or deterministic direction back into the way in which evolutionary biologists approach their day-to-day analyses of problems. This does not preclude discussions of more philosophical issues such as the role of prime movers in the origins of life, but these are currently beyond scientific resolution.

One of the least acceptable words in the evolutionary lexicon must be *purpose*. At one level, it resonates with teleological baggage, and, at another, it recalls the simplicity of the adaptive just-so stories that so enraged Lewontin and Gould (Gould and Lewontin 1979). However, the synonyms and words associated with *purpose* found in dictionaries are *intent, aim, design, function, role*, and *use*. In that range of associations can be found many of the major issues in evolutionary theory, and so it will be used here to develop an understanding of what might, or might not, be deeper structures in biology.

So, rather than abandoning the issue, there are undoubtedly some grounds for considering not purpose itself, but the issues in evolutionary biology where one is tempted to see a purposive element. In other words, leaving aside those theologically minded biologists who were determined to see the hand of God in nature, it is clear that the reason many were drawn towards a consideration of purpose in biology is that so much of it does, as Dawkins has argued (Dawkins 1987), give the appearance of there being a purpose. The illusion of purpose is the biological challenge, not purpose itself. The point of this approach is that, if we set about trying to explain those aspects of nature that appear purposive in terms of normal

Darwinian mechanisms, then we will achieve one of two aims. Either we will succeed and thus be able to account for this illusion without recourse to other than simple mechanisms. Or else we will fail and then perhaps have grounds to make a more radical reappraisal of the mechanisms of evolution.

Another way of putting this is—how can we have "purpose" in biology without "purpose"? The idea here is to use the metaphor of purpose to explore the evolutionary process and the patterns of life that are observable. To consider this issue, I will focus on what I think to be some of the most difficult aspects of evolutionary biology when it comes to simple Darwinian mechanisms, the ones that show the highest level of complexity, and see whether they give grounds for something more than just selection, chance, and adaptation. These "challenging cases" will be drawn from the field of human evolution; for, if there is any unifying theme in the more teleological approaches to evolution, it is that humans are indeed a special case.

We can start to identify where the temptation to look for purpose comes from by considering what in biology gives the illusion of purpose. To my mind, it comes from a number of overlapping themes: the functionality of adaptation; time's arrow and the evolution of "interesting novelty"; the evolutionary dominance of humans; and the role of the human mind in "terminating" evolutionary processes. While this essay will not attempt to restore purpose to the evolutionary lexicon in a way that might be hoped for by some, it will be argued that the concept does allow us to explore some of the more conceptually difficult areas of evolutionary biology.

Convergence, Deep Structure, and Purpose

One may ask why issues such as purpose and deep structural laws should be returning to evolutionary biology as serious issues, long after the scalpel of the modern synthesis had excised these exostoses from the skeleton of Darwinism. At first sight, it might appear that the strongly mechanistic approach of molecular biology would make this unlikely. One plausible answer lies in the growing evidence for and interest in convergence in evolution (Conway Morris 2003). Elsewhere in this present book, the extent, nature, and implications of convergence are considered in far more detail. Suffice it to say here that convergence does seem to be rife at all levels of the biological world. Convergence is a challenge to evolutionary theory, taking us back to some older issues, because it raises the question of directionality. Long eschewed as a distraction from the local nature of adaptive change, the observation that the same changes recur across many lineages at different times indicates that some "directions" of change are more likely than others. Furthermore, it is only a short step from this probabilistic view of the direction evolution might take to the acceptance of some form of progressive change.

That convergence is common is not at issue; its explanation is, however. Two prime candidate theories have been proposed. One is that convergence is evidence for what might be called the limited scope of biological materials; favored by authors such as Stephen J. Gould, this explanation rests largely on the idea that phylogenetic, developmental, and physical constraints restrict the forms that biological systems can take, and, therefore, change tends to be strongly limited to what is developmentally possible (Gould 2003). This explanation has been considerably reinforced by the evidence for the deep conservation of regulatory genes across many lineages. However, there is also mounting evidence that the genome can be dynamic at this level. These same regulatory genes are capable of producing widely divergent phenotypes at a very fundamental level (Averof and Akam 1995), and convergence at the very basic level of biological systems, such as the evolution of multicellularity, can occur there through entirely different genetic means (Maynard-Smith and Szathmary 1997). An alternative view is that convergence is so common because the adaptive problems faced by organisms in the struggle to survive recur frequently, and so selection tends to favor the same solutions (Conway Morris 2003). In this case, it is not the nature of biological materials that is limited but the "imagination" of the selective forces.

It is almost certainly the case that both explanations are true, and that it is a question of which is the most frequent and under what circumstances each might apply. For example, the occurrence of "agriculture" in both humans and ants is most probably a case of "convergent selection," whereas "developmental convergence" would apply in cases of similarities in limb function across the vertebrates. Certainly, the overall pattern of hominin evolution seems to support an element of convergent selection (Foley 1998). The point about both explanations, however, is that they remain strongly rooted in mechanistic and Darwinian biology and so do not threaten the ultimately reductionist approach of evolutionary theory. Directionality, far from becoming a teleological problem, becomes one of organism-based process, environmental selection, and probability.

If the directionality implied by convergence is not a challenge to evolutionary theory but rather, as Conway Morris (2003) has argued, underpins the power of Darwinism, it might be a challenge to another mainstay of Darwin's formulation of evolutionary theory—the notion of divergence. Evolutionary change is classically thought to be the divergence of descendent forms from their ancestral ones. However, it is misleading to think of divergence as being the antithesis of convergence—rather, it is a special case of divergence. When lineage A converges on lineage B, it is, in fact, diverging from its own ancestor, which is more "distant" from lineage B. Generally speaking, of course, it is also occurring completely in the absence—in time, or in space, or both—of lineage B. It is only the inquisitive biologist who brings together the occurrence of both lineages and discovers the convergence. By and large, sympatric, synchronic convergence is extremely rare. It follows from this that convergence as such is not a process of evolution but a pattern arising from the power of divergent selection. As such, it cannot be used, at the level of mechanism, to throw light on deeper structure of biology and the "purpose" of evolution. What it does do is provide a natural laboratory for testing evolutionary hypotheses.

Adaptation: From Function to Purpose

Ultimately, there can be little doubt that what drives one to see purpose in evolution is that adaptations are so clearly purposive. The wings of a bird have the purpose of flying or more immediately, providing lift, and therefore allowing flying. When a bird flaps its wings, it clearly has the purpose of flying; therefore, the adaptation for flying involves not just the wings but the whole central nervous system and the animal's cognition. It was the same argument that led to Paley's inference about a designer of nature, and, obviously, it is this that has led to a close link between ideas of design and ideas of purpose. However, it is on the latter that I will focus.

We can see purpose in adaptations, and as adaptations are the product of selection—the core of the evolutionary process—we can perhaps infer that there is purpose in evolution at this level. This is relatively trivial, although it has not gone without its controversies. The best-known of these is over the difference between adaptation and exaptation. According to Gould and Vrba (Gould and Vrba 1983), it is only an adaptation if it evolved specifically for that *purpose;* if feathers evolved for a purpose other than flight (say, thermoregulation), then there would be no purpose in the evolution of flight itself, only an exaptation would have evolved. However, as evolution is a seamless process of co-option from the very start, the difference is seldom as important as is made out. Given that the evolutionary process is indeed blind in terms of intention, the difference between an adaptation and an exaptation is largely semantic.

There is, however, a more interesting element in the purposive nature of adaptation. The illusion of purpose in an adaptation comes from the fact that it has a function; the function, for example, of the eye is to process visual information. Take away that purpose, and the eye has no function—and indeed is often lost in evolution as a result. In organisms, however, each adaptation is strongly integrated with others. The eye, for example, is coadapted with the visual cortex and is, in turn, linked to all sorts of other activities in movement through a process of coordination. When an animal "sees" a prey, its brain responds by initiating a whole series of movements, often very complex and highly coordinated, leading to capture, processing, ingestion, and digestion. In this sense, the process of seeing can prime a whole series of other processes, often in an anticipatory manner that provides a strong impression at least of purpose. This coadaptation may often be in terms of the autonomic nervous systems, but it is also the case that much of it is also concerned with higher cognitive processes and even conscious thought. To take a primate example, when a baboon grooms another while hiding behind a rock so that it cannot be seen by another member of the group, it has a purpose. While it can be disputed whether the cognitive purpose is genuinely to deceive or whether it is a simpler means of avoiding aggression, nonetheless, it would be hard not to see the positioning of the activity as a form of purposive behavior (Byrne and Whiten 1986). Any owner of a dog would find it hard not to accept this level of purposive action among animals. This leads to the perspective that even the simplest adaptations are tied in to the motivation of the animal—to flee from something it sees because it is dangerous or to move toward it if it is attractive.

Here we can see that, at the individual level—and that is the level at which most people accept that evolutionary mechanisms work—purpose is an integral part of the evolutionary process, and so, in that sense, there has been strong selection at a very early stage of evolution for the ability of an animal to turn stimuli into action through the process of motivation—purposive action toward a desired goal. In other words, purpose in this case is genuine in the sense that animals have been selected for purposive behavior. However, we also know that, at the fundamental level of mechanism, this is purpose as an illusion because the mechanisms involved are actually all about small-scale feedback systems, with no overarching goal. This is a classic case of the blind watchmaker in operation, using a series of causally linked stimuli and responses to produce behaviors and actions that enhance survival. An animal without this level of purposive action would not survive long.

Humans present an interesting case in this respect because we have purposive behavior in spades—or at least the illusion of it. What characterizes human behavior is the extent to which the human mind directs it. There are two ways of looking at the evolution of this: one, that it is a local adaptation, specific to humans and their evolutionary history among the primates. The other is to argue that it is a constrained system, in which the only way in which complex adaptive integration can occur is through this form of "central control." Evidence from the other highly intelligent mammals (the cetaceans) would suggest that the latter is the case, for there is a high level of convergence (Marino 2002). However, the complex adaptive integration of insect societies might well be used to indicate that it is not the only way in which such an end might be achieved by selection (Franks et al. 2002).

In summary, there is clear evidence that the process of adaptation has led to purposive behavior through the conventional processes of natural selection. This produces the illusion of purpose in two ways. The first is that, as the pattern of evolution is the sum of what individuals do to survive, then purpose is an important part of it. The second is that it can be argued that, overall, animals whose adaptations are better integrated—and among mammals at least that can be said to mean under higher levels of cognitive control—will be favored, and so, over time, there should be a trend towards more clearly directed behavior. However, it remains, at the macrolevel, purely an illusion because it is the product of natural selection. What is interesting, however, is that, at that level, what is the key signal or correlate of purposive behavior is the integration of adaptations—an area only now being opened up to study through molecular biology and cognitive science.

Time's Arrow and the Evolution of Novelty: Progress Is Real

The apparent pattern of progressive or directional change among lineages is an important element in any argument about the processes of evolution and inferences about higher order systems. Progress was seen by some as the evidence for a directionality that could be interpreted as evidence for a purpose (Teilhard de Chardin 1955). Such arguments have been widely rejected (Gould 1996). Leaving aside a contemporary dislike for concepts of progress, the primary evidence against it lies in the fact that conservative forms do not disappear. There are obviously strong adaptive reasons for this conservatism in nature. However, it is also the case that evolution does throw up novelty—organisms and adaptations exist now which did not exist in the past. In this sense progress is a real feature of evolution, because there is a very observable way in which changes are directional (Ruse 1996). Brain size, for example, does increase through time among primates and hominins (Jerison 1973; Martin 1983).

This directionality tells us about the nature of selection. Adaptations are context specific, and so the appearance of new adaptations will change the context in which they occur. A 500-cubic-centimeter brain in a world of 400-cubic-centimeter brains is in a different context from one in a population where everyone has a 500-cubic-centimeter brain. This, in effect, is the relentless world of the "red queen," and only changing cost/benefit ratios will act as a brake on the process if selective pressures remain.

The Red Queen hypothesis, first developed by Van Valen (Van Valen 1973) but implicit in much earlier evolutionary thinking, proposes that selection is shaped primarily by interactions between species, populations, and individuals, each of which is undergoing evolution. In Van Valen's view, each evolutionary shift in one unit generally represents a deterioration in the environment of another, hence leading to changed and intensified selective pressures. Evolution, thus, becomes a process of coevolution or, in more extreme forms, an arms race.

Here we are at the heart of the illusion of purpose: directionality is the product of competitive coevolutionary processes occurring cumulatively, observable with the benefit of hindsight. However, it is nothing more than the way in which selection operates. What adds complexity to this problem is that it is clearly lineage specific, in that some lineages persist in this progressive change more rapidly or further than do others.

Ecological Competition, Evolutionary Trends, and the Red Queen

A key question would, therefore, be—are there any patterns as to which lineages are more prone to such Red Queen mechanisms? What are the correlates of this? Humans can arguably be called one of the more extreme lineages and can be used to explore the problem. This is an empirical issue and requires more analysis than can be provided here. However, it could be argued that one of the key factors is the interaction between ecology and the overall competitive framework. Ultimately, adaptations require energy, and it is only the ability to sustain the costs of an adaptation ecologically that can permit continued evolutionary change. The evolutionary history of the hominines, which we are fortunate enough to be able to observe in more detail than in most lineages, shows that relatively major foraging changes have occurred throughout: meat-eating among early hominines, food processing through fire and technology, and ultimately food production.

We can explore the empirical basis for this in a little more detail. The earliest hominines, those existing in the Pliocene (5 to 1.6 million years), have brain sizes that are not significantly different from those of the living great apes, although they do show some minor relative increase (Klein 1999). The substantial changes in hominine brain size occur with the appearance of true genus *Homo*–*Homo ergaster* at around 1.6 million years, or probably a bit earlier (Walker and Leakey 1993). This change in brain size is associated with changes in ecology, indicated both by a new functional morphology (full bipedalism, reduced dentition and gut) and by changes in behavior (enhanced technology and increased dependence upon meat). The suggestion can be made that the increased access to high-protein food sources provided the increased energy necessary to fuel the growth and maintenance of a large brain (Foley and Lee 1991). A second phase of accelerated increase in the size of the human brain occurs in the last million years, especially between 350,000 and 200,000 years. In this case, it can also be argued that there is clear evidence in the archaeological record for a more sophisticated technology for hunting and a wider range of animals captured (Foley and Lahr 2003; Stiner et al. 1999). I would propose that the change in cognition inferred from this increase in brain size, whatever the selective pressures that may have underlain it, is fueled by ecological changes that enhanced the availability of energy. Although at a different scale, it can also be suggested that the major changes in human demography and distribution that occurred in the last ten thousand years, and that represent the period when humans became truly dominant as a species, are the product of the ecological changes brought about by the development of domestication and agriculture. This last example is discussed in more detail below.

In summary, directional change is an undoubted feature of evolution, giving the illusion of purpose to some, progress to others. I would argue, though, that this is explicable in terms of the way in which novel adaptations change the selective environment in a Red Queen manner. It is the specifics of ecology that will determine the extent of this process, leading to differential lineage patterns. At the moment, we are far from being able to specify the conditions that promote or inhibit such directionality. The study of convergence is obviously one approach to this problem (Conway Morris 2003), but this will require being able to specify the conditions of lineage evolution at a very general level to be able to carry out comparisons. The other is to examine the role of complexity in this: an initial position would be that this cumulative competitive process is inherently biased towards increased complexity and that is the primary effect of the way in which adaptive strategies change the context in which selection is occurring. Complexity may, in these Red-Queen-driven circumstances, be inevitable.

Human Evolution: The Empirical Basis for Dominance

At the public level at least, humans dominate our view of evolution. Humans are so radically different from other species and exert such an influence on the biosphere that it is all too tempting to see them as "further along" than other species, and, of course, it was this all-pervading notion of human superiority that led so easily to the view that humans are the "point" of evolution. Few, if any, evolutionary biologists would subscribe to this view, and some have gone so far as to argue that humans are simply an accidental exaptation of the evolutionary process.

The preceding discussion about the role of continuous coevolutionary and coadaptive processes, with the emphasis on the costs of continued change, would clearly undermine such arguments. It is, however, briefly worth considering when humans did become as distinctive and dominating as they now are. The short answer is very recently. If we apply the thought experiment of a visiting Martian and asked when such an alien would consider there to be a dominant species on the Earth, the answer is probably at two points. In terms of actual domination, from an ecological standpoint, where population numbers would be high and humans would have an effect on the environment, it is only the shift to agriculture in the last ten thousand years that would be such evidence. Prior to that, all the genetic, fossil, and archaeological evidence suggests that humans were small, scattered, and highly fluctuating populations, subject to local extinction (Foley and Lahr in press). Agriculture, not the evolution of the species, is the major ecological change. Indeed, I have argued elsewhere that this shift represents the most significant ecological change since the late Permian, with the evolution of major herbivores (Foley 1995). Ironically, agriculture is also essentially a major shift to herbivory among humans, a fact that may provide some insights into the nature of major biosphere transformations.

While the visiting Martian circling the planet would not notice anything significant about the human species until less than ten thousand years ago, if he or she (assuming sex is a universal of the evolutionary process) were to abduct a member of the species, then it would proba-bly find all the traits that make world domination possible from a cogni-tive and behavioral perspective at least two hundred thousand years ago. While an order of magnitude longer, it is still a very short period of time in the context of the history of life.

What do we learn from this about the issue of inevitability that is closely associated with the notions of purpose and progress? Perhaps the most important thing is that a detailed understanding of time and history is essential to unravel the "inevitability" of evolution. Viewed in general terms, there is a relentless trend towards the human end-point condition—one moment, in geological perspective, we are standing up, and the next we are landing on the moon. However, fine-grained analysis shows that there are large chronological gaps between events. The most intriguing of these is the gap between the emergence of the human species and its devel-opment of agriculture, the basis for major cultural change, intensification of urban life, and the development of the major technological changes that led to the modern world. That there are so many tens of thousands of years intervening between the appearance of modern humans and this demographic and ecological change—and millions since the emergence of the inception of the trend towards larger brains—suggests that there is nothing inevitable about the process except in terms of the interactive selective processes discussed earlier. More to the point, understanding why humans evolved can only be done by unraveling the detailed demographic and ecological history of the last few hundred thousand years. For there to be a gap of eight thousand to ten thousand generations implies com-plex selective pressures that could lead in different directions. Again, con-vergence may be important as a route into this problem (Conway Morris 2003), as it is clear that the path to human agriculture occurred more than once—although, intriguingly, at about the same time in different parts of the world.

The Human Mind and the End of Evolution

The final "hard case" that human evolution brings to the courtroom of evolutionary argument is whether evolution has changed as a result of the emergence of humans. It is a standard anthropological argument that the evolution of the conscious mind, language, symbolic thought, culture— select your favored trait—represented the emergence of a new evolutionary mechanism, which either supersedes or runs in parallel to conventional gene-based Darwinism, either way opening up new possibilities. If humans represent the end point—or one end point, at least—of evolution, then, in one sense, this could be seen as irreversible directionality in evolution.

What is the basis for this argument? Certainly, most of the world's population today are still relentlessly under some form of selection, through disease and malnutrition, and differential reproductive success is still a strong factor. The best we can argue is that selective pressures have changed and that their intensity is variable across populations. We can only speculate on what novel selective pressures are coming in with modernization that may affect global genetic structure (although we can say that the enormous size of the human population is likely to have massive inertial effects).

In that context, the evolution of the human mind does not necessarily change the mechanisms of evolution, and there is no evidence that the Darwinian process is at an end—there is no end point in sight. What is the case is that the presence of the human mind has radically changed what is selected, not just among humans but across the biosphere as a whole. However, humans are doing things that are different—notably the rate of change, the range of novelties introduced, and the scale of consequences. The detailed mechanisms of cultural evolution, such as have been developed by Boyd, Richerson, and others (Boyd and Richerson 1985; Cavalli-Sforza and Feldman 1981), have gone some way to showing how this has occurred and have provided some principles for understanding the patterns—all incidentally, rooted in the Darwinian process of individual selection. What I would argue, though, is that what is really happening here relates to the idea of constraints that are extensively used in evolutionary biology.

Constraints are poorly defined, although the concept broadly covers phylogeny, development, and physics (McKitrick 1993)—in my view, they are little more than things that selection has not yet overcome, a view that may be finding some empirical support (Beldade et al. 2002). Clearly, some may never be overcome—the effects of gravity, etc. While, in some respect, evolution over time does become more and more constrained— changing complex adaptations is too costly—there is no reason that this aspect should apply universally to all constraints. Some may indeed become more relaxed. For example, the mechanical constraints on the size of the human brain at birth, set by the size of the pelvis, may be removed either by the universal application of Caesarian sections or by ending the need for women to be able to walk, hence allowing larger pelves. Neither, however, is likely in the foreseeable future.

We can think about the way constraints can change theoretically. We can imagine any organism as being shaped by selection, and that these selective processes are subject to a myriad of constraints. These constraints can be seen as that organism's "population" of constraints. With increased "complexity" and the need for adaptive integration, then, over time, those constraints should increase. A more complex organism is probably subject to more constraints than a simple one. However, the number of constraints is not the only property they may have; they can also be considered to have a "resistance" value—that is, a measure of how hard it will be for them to be overcome. Another way of putting this is to see their resistance value as the costs of change. Some will have very high costs, others lower ones. As an organism evolves, it will change these resistance values; in many cases, the demands of integrative adaptation will increase the costs, but, in other, cases they may reduce them such that the constraints disappear—are, in other words, lost from the population of constraints. The point here is that constraints are neither equal nor constant, that they are context-specific, and, thought of probabilistically, some will, perhaps inevitably be removed even if others become more significant. What can perhaps be argued—speculatively, at least—is that the evolution of the human brain, greatly extended in capacity and capability over that of any other species—has through its very size and complexity removed various constraints inherent in the brains of other species (reduced the costs of overcoming those constraints), although, looking at the human organism as a whole, this cerebral evolution has also imposed a large number of other constraints, particularly in terms of life history.

To sum up this section, while human evolution clearly extends the evolutionary space very considerably, it is not necessarily the case that the basic rules of the game are changed—more that some applications of those rules become stricter (more constrained) and a few less so.

Conclusion

In this brief outline I have tried to use human evolution as a framework for considering the major issues of the evolutionary process and the questions of how these relate to the illusion of purpose, design, and progress. Humans are central to such debates because they are such outliers in comparison to the rest of the biological world. Their outlier status, I would argue, derives primarily from the extent of their evolution of purposive behavior, the way in which they clearly extend trends seen in the primates and mammals more generally, their domination of the biosphere, and their intelligence. The human species represents the hard case that is needed to test general models of the evolutionary process.

I have argued, in line with many evolutionary biologists, that there is no purpose in a fundamentally causative manner in evolution but that the processes of selection and adaptation give the illusion of purpose through the utter functionality and designed nature of the biological world. I have also argued that the illusion of purpose is further enhanced by the fact that selection has clearly favored strongly motivated behavioral abilities and that, in that sense, purpose resides in the genome and phenotype of organisms. These are general arguments that would apply to all living organisms, but I also argued strongly that this trait is both more enhanced in more complex organisms and is itself a spur to further complexity. I have tried to show that human evolution, the most challenging of all aspects of evolutionary biology because humans are so different in many ways from other animals, does not represent a departure from these general principles.

Looked at superficially, there is much to see in humans that might be seen as revolutionary in terms of process as well as achievement. However, I have argued that this is largely unwarranted at the most basic level of evolutionary mechanisms. Rather, the adaptive process that is driven by selection does have some law-like properties that may well—under the right circumstances (i.e., not for all lineages, for that is the local nature of evolution)—lead to more purposive behavior as a means of increasing or coping with complex adaptive integration and greater complexity and lead to constrained directional trends. These characteristics can be said to give evolution a repetitive and, hence, to some extent, inevitable pattern. Detailed study of human evolution, though, is likely to show that the underlying mechanisms remain stubbornly Darwinian.

The final conclusion I would draw is that evolution on other planets—or a rerun of evolution on this one—will lead to many similarities because of the law-like nature of these processes. However, the caveat I would add is that, while the human end point is a stable equilibrium outcome (so far!) of the evolutionary process, we should only perhaps conclude that it is the most probable outcome in a greater set of other possible outcomes. In a distribution of intelligences in the universe, on a sample of one, we might speculate that conscious, purpose-driven intelligence represents the mode.

Acknowledgments

Comments and discussion on the ideas expressed here from Marta Mirazon Lahr, Lucio Vinicius, and Simon Conway Morris have been invaluable. I am extremely grateful to Simon Conway Morris and the officers of the Templeton Foundation both for the invitation to contribute to the meeting and their support and hospitality. I would also like to thank George Coyne of the Vatican Observatory, Castel Gandolfo, not only for being so welcoming but also for allowing us to view so many of the first editions in the Vatican Library, books that changed the way we see the world.

References

Averof, M. and M. Akam. 1995. Hox genes and the diversification of insect and crustacean body plans. *Nature* 376: 420–23.
Beldade, P., K. Koops, and P. M. Brakefield. 2002. Developmental constraints versus flexibility in morphological evolution. *Nature* 416: 844–47.
Boyd, R., and P. Richerson. 1985. *Culture and the evolutionary process.* Chicago: University of Chicago Press.
Byrne, R., and A. Whiten. 1986. *Machiavellian intelligence.* Oxford: Clarendon.
Cavalli-Sforza, L., and M. Feldman. 1981. *Cultural transmission and evolution.* Princeton, N.J.: Princeton University Press.

Conway Morris, S. 2003. *Life's solutions: Inevitable humans in a lonely universe.* Cambridge: Cambridge University Press.

Dawkins, R. 1987. *The blind watchmaker.* Harlow: Longman.

Foley, R., and M. M. Lahr. 2003. On stony ground: Lithic technology, human evolution, and the emergence of culture. *Evolutionary Anthropology* 12: 109–22.

Foley, R. A. 1995. The causes and consequences of human evolution. *Journal of the Royal Anthropological Society* 1: 67–86.

———. 1998. Pattern and process in hominid evolution. In *Structure and contingency: Evolutionary processes in life and human society,* ed. J. Bintliff, 31–42. London: Leicester University Press.

Foley, R. A., and M. M. Lahr. In press. Flux and fragility: Demographic models and the relationships between Eurasian and African Late Pleistocene hominins. In *Biogeography and Neanderthal evolution,* ed. C. Finlayson.

Foley, R. A., and P. C. Lee. 1991. Ecology and energetics of encephalization in hominid evolution. *Philosophical Transactions of the Royal Society, London Series B* 334: 223–32.

Franks, R., S. Pratt, E. Mallon, N. Britton, and D. Sumpter. 2002. Information flow, opinion polling and collective intelligence in house-hunting social insects. *Philosophical Transactions of the Royal Society: Biological Sciences* 357: 1567–83.

Gould, S. J. 1977. *Ever since Darwin: Reflections in natural history.* London: Penguin.

———. 1996. *Wonderful life: The Burgess Shale and the nature of history.* New York: W W Norton.

———. 2003. *The structure of evolutionary theory.* Cambridge, MA: Harvard University Press.

Gould, S. J., and R. Lewontin. 1979. The spandrels of San Marco and the panglossian paradigm: A critique of the adaptationist programme. *Proceedings of the Royal Society B* 205: 581–98.

Gould, S. J., and E. S. Vrba. 1983. Exaptation—a missing term in the science of form. *Paleobiology* 8: 4–15.

Jerison, H. J. 1973. *Evolution of the brain and intelligence.* New York: Academic Press.

Klein, R. G. 1999. *The human career.* Chicago: University of Chicago Press.

Marino, L. 2002. Convergence of complex cognitive abilities in cetaceans and primates. *Brain, Behavior and Evolution* 59: 21–32.

Martin, R. D. 1983. *Human brain evolution in an ecological context.* New York: American Museum of Natural History.

Maynard-Smith, J., and E. Szathmary. 1997. *The major transitions in evolution.* Oxford: Oxford University Press.

McKitrick, M. 1993. Phylogenetic constraint in evolutionary theory: Has it any explanatory power? *Annual Review of Systematics* 24: 307–30.

Ruse, M. 1996. *Monad to man: The concept of progress in evolutionary biology.* Cambridge, MA: Harvard University Press.

Stiner, M. C., N. D. Munro, T. A. Surovell, E. Tchernov, and O. Bar-Yosef. 1999. Paleolithic population growth pulses evidenced by small animal exploitation. *Science* 283: 190–94.

Teilhard de Chardin, P. 1955. *Le phenomene humain.* Paris.

Van Valen, L. 1973. A new evolutionary law. *Evolutionary Theory* 1: 1–30.

Walker, A. C., and R. E. Leakey, eds. 1993. *The Nariokotome Skeleton.* Cambridge, MA: Harvard University Press.

10 PURPOSE IN A DARWINIAN WORLD

Michael Ruse

Darwinism appeared, and, under the guise of a foe, did the work of a friend.

Aubrey Moore

Rival Visions

Erasmus Darwin loved to eat. He would come to lunch and wolf down fish, fowl, and flesh, followed by rich desserts and lashings of cream, happy in the knowledge that, but a few hours later, he could start all over again. In the grand tradition of Saint Thomas Aquinas, his own table was cut in a hollow semi-circle so that he could more easily reach the provisions laid out before him. Yet expensive and onerous though it may have been to entertain him, Erasmus Darwin was a much-sought-after guest and dined out all over the British Midlands where, toward the end of the eighteenth century, he made his home. He was not only a brilliant doctor— George III begged him many times to come south and to take on the role of court physician—but he was an inspired conversationalist, with ideas and enthusiasm bubbling forth and captivating listeners. A member of the Lunar Society, a group of scientists and manufacturers eager to harness nature's forces for the ends of industry, Darwin had thoughts on agriculture and on politics (he was a close friend of Benjamin Franklin), on school teaching (he wrote a treatise on the education of women), on the Linnean sexual system of classification, and on much, much more—thoughts that he hymned in verse, for Erasmus Darwin was also one of the best-known poets of his age.

Above all, Erasmus Darwin—the grandfather of Charles Darwin—was an enthusiast for progress, the belief or ideology that humans unaided can improve both theoretical knowledge and social conditions, if only they work long enough and hard enough. Like many of his contemporaries, Erasmus Darwin was a deist, believing in a God who had created and who then let the unguided laws of nature do everything subsequent. It is our task here on Earth to discover and to use these laws. Instead of letting nature take its unguided course, it is for us to work to improve and increase—improve our understanding and increase the bounties of our way of living. Reflecting this upward thrust was evolution. For Erasmus Darwin, the ongoing transmutation of organisms was not something to be proven or to be added on. It was part and parcel of his world picture. We go from the blob to the human, from the monad to the man, from (as he himself said) the monarch (the butterfly) to the monarch (the king).

> Imperious man, who rules the bestial crowd,
> Of language, reason, and reflection proud,
> With brow erect who scorns this earthy sod,
> And styles himself the image of his God;
> Arose from rudiments of form and sense,
> An embryo point, or microscopic ens!
>
> Darwin 1803, 1, 295–314

This is a world of exuberant life in every sense. Just as (to use a favorite metaphor of the time) a world that produces goods by machine is superior to a world that produces goods by hand, so also a world that produces organisms through law is superior to a world that produces organisms through miracle. Thus, never make the mistake of thinking that the God of Erasmus Darwin is dead or uninterested. The deity cares desperately about his creation. It is rather that it is now for us humans to continue the work of development and improvement. For this reason, the world is one of purposes, of intentions, of ends. But rather than waiting passively for these to be realized, we humans must join with God in achieving these aims (Ruse 1996).

Writing at the same time as Erasmus Darwin, and also living in the British Midlands, was another man with a very different agenda. William Paley, archdeacon of Carlisle, was wracked with pain (he probably had cancer of the bladder) and too sick to minister to his congregation. He could find relief only in the world of the mind, and having written what was generally considered the standard work on Christianity and its empirical proofs—*View of the Evidences of Christianity* (1794)—by century's end, he was turning to a new task. He too was interested in purpose and intention, although in a way different from Darwin. Paley wanted to refurbish and give the definitive account of the most famous argument for the existence of God. He wanted to offer a detailed and convincing positive account of the argument from design or the teleological argument: the argument that began with Plato (perhaps even Socrates) and was made famous by Saint Thomas Aquinas, that drew attention to the design-like nature of the living world, and that concluded that this can be no chance but must be the completed intention of a loving and all-powerful deity.

In his *Natural Theology* (first edition, 1802), in what has become one of the most famous of all passages cropping up in undergraduate curricula—from then until now—Paley drew a contrast between a rock and a watch.

In crossing a heath suppose I pitched my foot against a stone, and were asked how the stone came to be there, I might possibly answer, that for any thing I knew to the contrary it had lain there for ever; nor would it, perhaps, be very easy to show the absurdity of this answer. But supposing I had found a watch upon the ground, and it should inquired how the watch happened to be in that place, I should hardly think of the answer which I had before given, that for any thing I knew the watch might have always been there.

Paley 1819, 1

A watch demands a watchmaker. Likewise, the eye—in so many respects analogous to a human artifact like a telescope—demands an eye maker. For Paley and his audience, this could be none other than the Great Optician in the Sky. Things so wonderful and complex as the hand and the eye do not happen by chance. There must be a deity responsible, a being that Paley and his readers happily identified with the God of Christianity, a God whose creation that was no less filled with purpose and meaning than the God of Erasmus Darwin. It was just that, whereas the purpose and meaning for Erasmus Darwin's God came from progress—

what he had done in the world of organisms and what we are doing and are to do in the world of knowledge and culture and physical being—the purpose and meaning for William Paley's God came from direct divine intention—the ways in which he had miraculously designed and created organisms, inhabitants of a world in which the Christian drama of sin and redemption plays itself out (Ruse 2003).

Charles Darwin and the "Origin of Species"

It is easy and natural to think that Darwin and Paley represent rival visions, one looking forward and the other looking backward: two understandings of purpose, one for the future and one for the past. Erasmus Darwin is the man of the future, with progress and science and industry and evolution, backed by a God for the new and anticipated age. William Paley is the man of the past, with providence and faith and design, backed by a God of the old and forgotten age. There is some truth in this. The nineteenth century was the Age of Progress—of evolution also, thanks to Charles Darwin and his many supporters like Thomas Henry Huxley in Britain and Asa Gray in America. Many were happy to tie progress and evolution together in a tight synthesis. The very popular writer Herbert Spencer thought of progress as a move from the undifferentiated to the differentiated or as what he called a move from the homogeneous to the heterogeneous:

Now we propose in the first place to show, that this law of organic progress is the law of all progress. Whether it be in the development of the Earth, in the development of Life upon its surface, in the development of Society, of Government, of Manufactures, of Commerce, of Language, Literature, Science, Art, this same evolution of the simple into the complex, through successive differentiations, hold throughout. From the earliest traceable cosmical changes down to the latest results of civilization, we shall find that the transformation of the homogeneous into the heterogeneous, is that in which Progress essentially consists.

Spencer 1857, 244–45

Together with this "progressphilia," many Christians dismissed Paley as outdated. We must all move with the times. Far from seeing progress as opposed by providence, they came to see the two as essentially one (Ruse 2005). The charismatic American preacher Henry Ward Beecher wrote:

"If single acts would evince design, how much more a vast universe, that by inherent laws gradually builded itself, and then created its own plants and animals, a universe so adjusted that it left by the way the poorest things, and steadily wrought toward more complex, ingenious, and beautiful results!" He continued: "Who designed this mighty machine, created matter, gave to it its laws, and impressed upon it that tendency which has brought forth the almost infinite results on the globe, and wrought them into a perfect system? Design by wholesale is grander than design by retail" (Beecher 1885, 113). Others condemned visions of progress as anti-Christian but still wanted little truck with Paley and his arguments. The great John Henry Newman, convert from Anglicanism to Catholicism, wrote in 1870 (twenty-five years after he converted), in correspondence about his seminal philosophical work, *A Grammar of Assent*:

I have not insisted on the argument from design, because I am writing for the 19th century, by which, as represented by its philosophers, design is not admitted as proved. And to tell the truth, though I should not wish to preach on the subject, for 40 years I have been unable to see the logical force of the argument myself. I believe in design because I believe in God; not in a God because I see design.

Newman 1973, 25:97

This interpretation of history—praise for the vision of Erasmus Darwin and critique for the vision of William Paley—has to be a truncated picture, trimmed down to one of distortion. Focus in on the seminal work for our topic, the work that established the fact of evolution as a given and reasonable hypothesis: *On the Origin of Species*, published in 1859 by Charles Darwin. No one had read Paley with more care than had Darwin, and no one agreed more fully that the design-like nature of the hand and the eye—what Darwin called adaptation or contrivance—is the defining and significant feature of living nature (Ruse 1999). Darwin's mechanism of natural selection—more are born than can survive and reproduce, and only the fittest get through to parent future generations—is intended to speak explicitly to the problem of design. Hands and eyes are not just chance features but are things that come into being precisely because they are design-like—design-like to enable their possessors to survive and reproduce.

This said, however—the fundamental significance for Charles Darwin of the immediate, purposeful nature of organic attributes—the *Origin* did not leave things as they had been. On the one hand, no longer was one thinking of design as something perfect and optimal. For the Darwinian, winning is what counts. Adaptations do not have to be the best possible; they simply have to be better than those of rivals. On the other hand, no longer was the inference to the deity obligatory. Complex contrivance has to have a cause. Before Darwin, appeal to God as the cause of adaptation was the only option. After Darwin, one could simply ascribe everything to the working of blind law. In the memorable words of today's most popular writer on evolution, Richard Dawkins, only after Darwin was it possible to be an intellectually fulfilled atheist. Dawkins notoriously has gone the route of nonbelief. Darwin did not make this obligatory. At the time of writing the *Origin*, like his grandfather Charles Darwin was a deist. But, from now on, a Christian accepting the power of selection would be advised to turn from the theology of Paley to that of Newman. In other words, although a Darwinian would think of the hand and the eye as having immediate purposes, one cannot read God's intentions from them. As a Christian, one should interpret them in terms of God's intentions, but this is another matter.

What about the other side to purpose: progress? As with adaptation, there is no simple connection between the evolutionism of the *Origin* and doctrine of progress. Darwin himself was an ardent progressionist, in thinking about human society and culture as well as in thinking about biology. But he was ever wary of crude and enthusiastic endorsements of biological progress, both because he knew of many exceptions and also because he saw that progress in any absolute sense implies a kind of overall purpose and value to the world that modern science tries to exclude. Methodologically, the scientist strives to be an atheist, even if he or she accepts a fuller and more meaningful metaphysical picture of ultimate reality. Hence, just as in accepting Paley's emphasis on the design-like nature of adaptation, Darwin nevertheless knocked it sideways, so also in accepting his grandfather's emphasis on progress, Darwin nevertheless knocked that sideways. This was bound to be, for, as Darwin saw clearly, the immediate purpose of adaptation and the long-term historical purpose of progress have to be connected, if they are not, indeed, one. Immediate purpose builds through history to become ever better and more complex and satisfying, thus adding up to long-term purpose. Adaptation leads to progress. Patches of skin become light-sensitive become full-blown eyes.

The adaptation–progress connection, as seen by Darwin, came through what today's evolutionists label "arms races." Organisms or lines of organisms compete against each other, and, in the long run, what happens is that one gets improvement, overall. This cannot be improvement on some absolute scale, for, ultimately, it is built on a relative notion of adaptation—organic features are never perfect: they are just better than those of others. Likewise with progress. For Darwin himself, however, this relative merit was enough, for he thought that the end point of selection-driven evolution had had to be humans and their brains and consciousness. And what more could a good Victorian demand?

If we take as the standard of high organisation, the amount of differentiation and specialisation of the several organs in each being when adult (and this will include the advancement of the brain for intellectual purposes), natural selection clearly leads toward this standard: for all physiologists admit that the specialisation of organs, inasmuch as in this state they perform their functions better, is an advantage to each being; and hence the accumulation of variations tending toward specialisation is within the scope of natural selection.

Darwin 1859, 222

Life's Solution?

The world of the Darwinian is deeply purposeful, if not in an entirely traditional way. As with adaptation, one cannot read from biology any absolute or definitive notions of progress—monad to man. But if one comes to the world wanting to read in meaning and progress, then it is certainly open for one to do so. And this includes the Christian, especially the one who agrees with people like Beecher that, far from progress being antithetical to the Christian vision, it can complement and enrich it. To continue the passage quoted at the beginning by the late-Victorian, Anglo-Catholic, Oxonian theologian Aubrey Moore, thanks to Darwin, "We must frankly return to the Christian view of direct Divine agency, the immanence of Divine power from end to end, the belief in a God in Whom not only we, but all things have their being, or we must banish him altogether" (Moore 1890, 268–69).

Does the world of the Darwinian speak adequately of purpose today? Recognize that Darwinism considered as science has matured since the days of the *Origin*—most particularly thanks to the infusion of adequate theories of heredity (first Mendelian and later molecular genetics)—and, in order to search for a positive case, ignore the fact that many today are still not convinced that natural selection is an entirely adequate cause of evolutionary change. Ask whether someone who accepts evolution through natural selection can speak positively of adaptation and progress, in a way that can satisfy not just scientifically but, in a broader sense, ultimately a sense that captures a religious yearning for purpose? Some would deny that it can. The late Stephen Jay Gould, notoriously, was not enthused by notions of adaptation and even less by thoughts of progress. He pronounced it an illusion and not a very nice one at that. He spoke of the idea as "a noxious, culturally embedded, untestable, nonoperational, intractable idea that must be replaced if we wish to understand the patterns of history" (Gould 1988, 319). Gould was a major booster of the contingency of life. Any thought that we might be the favored children of God, that we might be *the* (or a significant) reason for the creation, is simply hubris.

Since dinosaurs were not moving toward markedly larger brains, and since such a prospect may lie outside the capabilities of reptilian design . . . , we must assume that consciousness would not have evolved on our planet if a cosmic catastrophe had not claimed the dinosaurs as victims. In an entirely literal sense, we owe our existence, as large and reasoning mammals, to our lucky stars.

Gould 1989, 318

Is this pessimistic language the true voice of modern-day Darwinism? Clearly it is not if we are thinking in terms of adaptation. Today's Darwinians are as enthusiastic about the design-like nature of the organic world as was Darwin and Paley before him. George Williams, a man who jokes that the only good advice that he got from his priest was the direction to the front door, writes:

Whenever I believe that an effect is produced as the function of an adaptation perfected by natural selection to serve that function, I will use terms appropriate to human artifice and conscious design. The designation of something as the means or mechanism for a certain goal or function or purpose will imply that the machinery involved was fashioned by selection for the goal attributed to it.

Williams 1966, 9

But what about long-term purpose, and what about progress? Recently, the Cambridge paleontologist Simon Conway Morris has made a vigorous, Darwin-inspired attempt to refurbish a sense of progress, based on a selection-driven concept of adaptation. Conway Morris' basic starting position is that only certain areas of potential morphological space are going to be capable of supporting functional life, and to this he adds the assumption that selection is forever pressing organisms to look for such potential, functional spaces. Hence, if such spaces exist, sooner or later, they will be occupied—probably sooner rather than later and probably many times. Conway Morris draws attention to the way in which life's history shows an incredible number of instances of convergence—instances where the same adaptive morphological space has been occupied again and again. The most dramatic perhaps is that of saber-toothed, tiger-like organisms, where the North American placental mammals (real cats) were matched item for item by South American marsupials (thylacosmilids). Clearly existing was a niche for organisms that were predators, with cat-like abilities and shearing/stabbing-like weapons, and natural selection found more than one way to enter it. Indeed, it has been suggested that, long before the mammals, the dinosaurs might also have found this niche.

Conway Morris argues that this sort of thing happens over and over again, showing that the historical course of nature is not random but strongly selection-constrained along certain pathways and to certain destinations. From this, Conway Morris concludes that movement up the order of nature, the chain of being, is bound to happen, and eventually some kind of intelligent being (what has been termed a "humanoid") is bound to emerge. We know from our own existence that a kind of cultural adaptive niche exists—a niche where intelligence and social abilities are the defining features. More than this, we know that this niche is one to which other organisms have (with greater or lesser success) aspired. We know of the kinds of features (like eyes and ears and other sensory mechanisms) that have been used by organisms to enter new niches; we know that brains have increased as selection presses organisms to ever new and empty niches; and we know that, with this improved hardware, have come better patterns of behavior and so forth (more sophisticated software). Could this not all add up to something?

If brains can get big independently and provide a neural machine capable of handling a highly complex environment, then perhaps there are other parallels, other convergences that drive some groups toward complexity. Could the story of sensory perception be one clue that, given time, evolution will inevitably lead not only to the emergence of such properties as intelligence, but also to other complexities, such as, say, agriculture and culture, that we tend to regard as the prerogative of the human? We may be unique, but paradoxically those properties that define our uniqueness can still be inherent in the evolutionary process. In other words, if we humans had not evolved then something more-or-less identical would have emerged sooner or later.

Conway Morris 2003, 196

Like all powerful-but-*prima-facie*-simple ideas, there are depths here. So, let us pause to see what Conway Morris is claiming and doing and (as importantly) what he is not claiming and not doing.

Conway Morris is absolutely in the tradition of Charles Darwin himself. He is using natural selection and only natural selection as his significant causal motor. For Conway Morris, the key to understanding the history of life is adaptation, and this is caused by selection. He is postulating that there exist, independently of selection—of evolution, even— ecological niches that are waiting to be invaded by the right kinds of organisms. There is, for instance, dry land —a space that (in the case of animals) can be occupied only by organisms that can breathe on land, that can move about on land, and that can find food on land (or, at least, can go from land to find food, and then return). In fact, Conway Morris' position is a little more sophisticated than this. Niches are themselves subdivided into more specialized areas. Dry land, for instance, would have open niches on plains, woodland niches, jungle niches, desert niches, and so forth. Organisms might share a major niche but then move to different subniches. Humans and oak trees share the dry-land niche, but they do not share the intelligence niche. Humans are in the intelligence niche. Oak trees are in the photosynthesis niche.

Some critics of Darwinism have argued that organisms create niches as much as they discover them, but this is not really Conway Morris' position. Niches exist in their own right, and it is possible—and, at times, actual— that organisms can get into them by different routes: the placental and marsupial saber-toothed tigers, for instance. The reason that organisms get into (or fail to get into) niches is pure Darwinism. There is a constant pressure to survive and reproduce, and organisms are looking always for opportunities to succeed. Moving to a new niche reduces competition from others and gives new opportunities hitherto unavailable. To make a way into the niche, one needs appropriate adaptations—lungs for breathing, for instance—and these come through selection brought on by struggle. Although Conway Morris sometimes talks of constraints, his constraints are purely adaptive. If you have lungs, then you can breathe, and you can live on land. If you do not have lungs, then you cannot live on land (unless you have alternative equivalent adaptations, as do oak trees).

Because these niches are essentially discovered rather than created, what Conway Morris does not do is use Darwinism—or anything else, for that matter—to make the niches themselves, at least not to make the major or basic niches. They are provided. Just as Darwinism does not explain the existence of oxygen and hydrogen and the bonds between them that make possible water—although it certainly makes much of the way in which organisms can utilize water—so Darwinism does not explain the niches as such—although it certainly makes much of the way in which organisms can utilize the niches. Hence, if we think roughly of water, land, air, and culture as a series of basic niches, they are givens. As one gets down to the subniches, then Darwinism does start to come into play more and more. Just as Darwinism might explain why human physiology uses water rather than gasoline, so Darwinism might explain how jungle canopies open up niches for birds and insects and so forth.

As Darwinism does not explain the existence of niches, so likewise Darwinism does not really explain any ordering of the niches. It is true that water probably had to come before land and air, but whether land organisms had to come before air organisms seems a debatable point. In any case, any value judgments of an absolute kind are as barred to Conway Morris as they were to Darwin. Land organisms are at least as sophisticated as airborne organisms, and marine mammals have special adaptations that equal or better any adaptations of their land-based mammalian cousins. That is not to deny that those organisms that have made the breakthrough to culture are very complex and sophisticated. Nor is it to deny that we might want to judge them better or more valuable—I for one would think very peculiar any human who did not want to make this judgment. But, as with any Darwinian scenario, this is a judgment that we may make rather than one that we extract from the science. In fact, one might say that, biologically speaking, certain diseases—HIV and influenza, for instance—stand a chance of doing better than humans. They are certainly doing better than our close relatives, the great apes.

Does Darwinism give any reason to think that Conway Morris' progress (let us now use this term) is in some sense necessary—that it is something that selection makes probable? As the title of his book proclaims—*Inevitable Humans*—Conway Morris certainly thinks that this is so. And in a sense, no one is going to disagree. Even Stephen Jay Gould allows some kind of progress simulation. He argues that life's history is like a drunken man on a sidewalk, bounded on one side by a wall and the other by a ditch. Eventually, the man will fall into the ditch because he cannot go through the wall and his random path will take him to and over the other edge. So similarly, simple organisms cannot get simpler, but they can get more complex, ending ultimately in intelligence. But Conway Morris argues for something stronger, and speaking as a Darwinian, he is surely justified. Selection is keeping up a pressure to invade new niches, and, although entry is never inevitable, over time one expects such niches to be invaded—more than once, as the saber-tooth example shows. To use a sporting metaphor, it is not inevitable that the New York Yankees be perennial challengers for the world championship of baseball, but, given their payroll, it is hardly surprising that they are.

Implications for Purpose

Returning the direct focus now to the notion of purpose, conclude by seeing what more juice—scientific and theological—can be extracted from this discussion. Taking the science first, if in fact we do have a progression of niches—water, land, air, culture—why should they stop with culture and intelligence? Why not move on at least to another basic niche—why not move on to an infinity of such niches? You might say that you cannot imagine what a further niche would be like. But this is no argument to a Darwinian. Natural selection has made us able to deal with our circumstances—getting out of the jungle and on to the plains, for a start. There was neither need nor obligation to give us the powers to peer into the ultimate mysteries of creation. In fact, being too philosophical can have its downside. It makes for worry and doubt and indecision.

Is there reason to think that there is more than we can comprehend fully? Modern science makes us modest in this respect. Think of quantum mechanics, for instance. In the words of Richard Dawkins, "Modern physics teaches us that there is more to truth than meets the eye; or than meets the all too limited human mind, evolved as it was to cope with medium-sized objects moving at medium speeds through medium distances in Africa" (2003, 19). Should we think of the next niche up as the world of super-intelligence, perhaps the sort of dimension reported on by mystics? Well, you can if you want, but speaking as a Darwinian, I do not find this terribly helpful or insightful. The whole point is that it is a dimension of which we are ignorant. It will not be so much a world where the laws of nature as we know them—logic and mathematics, too—are broken. It will certainly not be a world to bring comfort to anti-scientists like the creationists. It will simply be a world beyond our ken. If you think that culture is a kind of super-air or -land, then perhaps the next niche would be a kind of super-intelligence. But it is something very different from regular intelligence, and that is about all we can say.

What we can say is that we expect some kind of physical continuity. The move from water to land to culture did not mean the end of DNA or cells or proteins or bodies. Whatever form the next niche might take and whatever might be the needed adaptations for entry, one would not expect future beings to have lost their corporeal nature. It might perhaps be that everyone or thing could have moved to a purely spiritual dimension— rather like disembodied Platonic forms or the ghostly undead of gothic novels—but Darwinism expects otherwise. If, as many think, Darwinism operates in the realm of culture itself—ideas struggling against ideas and machines against machines—it might possibly be that future beings would no longer be based on organic materials, as is the case for us. But speculations like these take us from the realm of science to fiction. What might be plausible is that the future beings in some sense lose personal identity, at least to the extent of becoming part of a unified whole. We certainly see how the ant colony works more efficiently as a kind of collective organism, with individuals as parts rather than as separate entities.

Finally, still at the level of science, the Darwinian is less than optimistic about ever actually achieving this future higher niche—at least, about our descendants ever reaching the niche. Part of the reason for pessimism is that one occupant at the top of the intelligence niche seems to discourage competitors. Look at the story of the Neanderthals. Look at the present state of the great apes, a state for which we humans are primarily responsible. If, unbeknown to us, a superior niche is somewhere in the universe already occupied, then it could well be that these inhabitants would not encourage newcomers. Or they might want to encourage newcomers but by their very existence discourage them. Part of the reason for pessimism is that the present state of human evolution gives no cause for thinking that we are on the way to something higher. Brain size, in fact, has gone down since the Neanderthals, and, given modern medicine and so forth protecting us from nature's forces, there is little reason to think that things are going in a way that we might describe as progressive. At most, we have people in the third world inadvertently being selected to put up with poverty and disease and so forth. I am not now saying that modern medicine is a bad thing or that third-world poverty and disease are good things. I am saying that they are not things that seem to lead toward entering bigger and better niches.

Part of the reason for pessimism is that human intelligence has evolved because of, and in such a way as to deal with, other humans. We are a highly social species, and our intelligence evolved in major part to make us social and to keep us this way. But, even if it works reasonably well in a biological sense for now, it is not perfect. Along with our adaptations for being social—helping others and so forth—we have a dark side, that side was equally produced by selection. We are selfish and suspicious of others and tend toward violence when we think it meets our needs or ends. And, because time is money, we tend to favor quick solutions that work for the short run, even despite difficulties down the road. Although Richard Dawkins has said truly that one should never underestimate the power of natural selection—as soon as one declares something impossible, one finds that nature has done it—it does seem that intelligence, the product of working with others, will generally have this dark side. And this being so, not just here but generally in time and space (that is, throughout the universe), one would expect that denizens of a niche like ours would have the intelligence to find weapons of great destruction and lack the ability to proscribe their use forever, through eternity. In our case, does anyone truly think that, in the next (say) twenty thousand years, no one anywhere will detonate a nuclear weapon? My point is that our niche of intelligence and culture may be self-limiting. Hope of ever going on to another niche, although physically possible, may be biologically barred. Natural selection can take us a long way but then turns back on itself and bars further advance.

What does this have to do with theological issues, and how does this tie in with purpose? Let it be stressed again that, since Darwin, issues of proof or confirmation are not even on the table. What has just been said barely proves anything scientific, let alone anything theological. In the language of theologian Wolfhart Pannenberg, at best we can hope for a theology of nature rather than a natural theology. However, the Darwinian scenario does resonate strongly with the Christian picture and perhaps with other religions also. It affirms that we are much but that we are limited. There may be dimensions of reality beyond our grasp. Christianity has always affirmed both our abilities and the ultimate mystery of being. In the words of John Paul II, "knowledge refers back constantly to the mystery of God which the human mind cannot exhaust but can only receive and embrace in faith. Between these two poles, reason has its own specific field in which it can enquire and understand, restricted only by its finiteness before the infinite mystery of God" (1998, 14). Could it not be that God and perhaps other beings occupy the higher niches and that we today can only see as through a glass darkly? There is an overall, purposeful progression to life, and we humans are—as in the old pictures of the chain of being—only part way up the path.

Second, if corporeal being is going to be as important as thinking or spiritual being, this fits not only with today's philosophical inclinations to some kind of monism—perhaps a neo-Spinozistic identity theory—but also with traditional Jewish thinking (as represented by Paul against the Greek-influenced Augustine) that the body has its role in the ultimate scheme of things. Whatever may be the case, Darwinism suggests that the body in some form will have its crucial role. Whether Christianity also favors some kind of oneness or integration of being, I will leave for others to speculate. It is a view that would certainly find favor with many mystics.

Third and finally, our inability to move naturally to higher dimensions also parallels many strands of Christian thought. We are not just limited but, in some way, necessarily limited by our dark side. In traditional language, we are tainted with original sin and can never expect to move forward and upwards purely on our own. This limitation does not mean that life is meaningless. If anything, it emphasizes the meaning of life. We must strive to overcome our failings, else we will lose what we (or nature) have already achieved. And more than this, we must recognize that we are not gods and that there is a higher dimension, on which we, unaided, will and can never achieve. There is a progress, a purpose to life, and we have our roles to play and our obligations to fulfill. We can never do it alone, but that is what the Christian has always said. And that is a good point on which to end this discussion. Darwinism has major implications for thoughts of purpose—implications that are deeply inspiring and no less deeply humbling.

References

Beecher, H. W. 1885. *Evolution and religion*. New York: Fords, Howard, and Hulbert.

Conway Morris, S. 2003. *Life's solution: Inevitable humans in a lonely universe*. Cambridge: Cambridge University Press.

Darwin, C. 1859. *On the origin of species*. London: John Murray.

Darwin, E. 1803. *The temple of nature*. London: J. Johnson.

Dawkins, R. 1986. *The blind watchmaker*. New York: Norton.

———. 1995. *A river out of Eden*. New York: Basic Books.

———. 2003. *A devil's chaplain: Reflections on hope, lies, science and love*. Boston and New York: Houghton Mifflin.

Depew, D. J., and B. H. Weber. 1994. *Darwinism evolving*. Cambridge, MA: MIT Press.

Gould, S.J. 1988. On replacing the idea of progress with an operational notion of directionality. In M. H. Nitecki, editor, *Evolutionary Progress*, ed. M. H. Nitecki, Chicago: University of Chicago Press. 319–38.

Gould, S. J. 1989. *Wonderful life*. New York: Norton.

John Paul II. 1998. *Fides et ratio: Encyclical letter of John Paul II to the Catholic bishops of the world*. Vatican City: L'Osservatore Romano.

McNeil, M. 1987. *Under the banner of science: Erasmus Darwin and his age*. Manchester, UK: Manchester University Press.

Moore, A. 1890. The Christian doctrine of God. In *Lux mundi*, ed. C. Gore, 41–81, London: John Murray.

Newman, J. H. 1973. *The letters and diaries of John Henry Newman*, vol. 25, ed. C. S. Dessain and T. Gornall. Oxford: Clarendon Press.

Ospovat, D. 1981. *The development of Darwin's theory: Natural history, natural theology, and natural selection, 1838–1859*. Cambridge: Cambridge University Press. Reissue 1995.

Paley, W. [1802] 1819. *Natural Theology*. Vol. 4 of *Collected works*. London: Rivington.

Pannenberg, W. 1993. *Toward a theology of nature*. Louisville, KY: Westminster/John Knox Press.

Richards, R. J. 1992. *The meaning of evolution: The morphological construction and ideological reconstruction of Darwin's theory*. Chicago: University of Chicago Press.

Ruse, M. 1996. *Monad to man: The concept of progress in evolutionary biology*. Cambridge, MA: Harvard University Press.

———. 1999. *The Darwinian revolution: Science red in tooth and claw*. Chicago: University of Chicago Press.

———. 2003. *Darwin and design: Does evolution have a purpose?* Cambridge, MA: Harvard University Press.

———. 2005. *The evolution–creation struggle*. Cambridge, MA: Harvard University Press.

Spencer, Herbert. 1857. Progress: Its law and cause. *Westminster Review* 67: 244–67.

Williams, George. 1966. *Adaptation and natural selection*. Princeton: Princeton University Press.

11 PLUMBING THE DEPTHS

*A Recovery of Natural Law and Natural Wisdom in the
Context of Debates about Evolutionary Purpose*

Celia Deane-Drummond

Theologians have wrestled long and hard with the implications
of the seeming purposelessness of evolution implicit in biological
theories, especially of the kind promoted by Stephen Jay Gould.
The assumption is that contingency, understood as aimlessness, is
the most important feature in the evolutionary trajectory and that
this contingency implies purposelessness, so that accommodation
with transcendent notions of purpose becomes virtually impossible. One way through this difficulty is to argue that contingency
is not necessarily synonymous with purposelessness, though most
biologists would argue that any idea of purpose is unnecessary in
order to understand the process of evolution. Some may go even
further and suggest that purpose is not simply unnecessary; it is
"illusory," created by human minds in order to serve an adaptive
function, in other words, it has survival value (Foley 2008). All
religious beliefs would, similarly, fall into this category of being
useful in evolutionary terms. However, while it might be possible to argue that psychological tendencies toward religious belief
have biological roots, the content of such beliefs cannot be fixed
through crude biological or even cultural determinism but operate at levels of understanding that are beyond simple biological
analysis. This is much the same as saying that evolutionary biology

cannot be "reduced" to physics or mathematics, even if physics or mathematics illuminates in a descriptive way something about the way evolutionary biology works. Simon Conway Morris' alternative evolutionary hypothesis that puts far greater weight on the phenomena of convergence points toward the possibility of a form of evolutionary "purpose," though more accurately it is perhaps more correct to speak of a "restrained contingency" (Conway Morris 2003, 2008). The intention of this chapter is not so much to enter into biological debates about which theory of evolution is the most reasonable but to engage from a theological point of view using natural law theory with Conway Morris' hypothesis.

It is important to point out at the outset what this chapter is not doing. It is not attempting to recover notions of "design" in the universe as a way of demonstrating the existence of God. These versions of natural theology have historically run aground on the twin horns of theological determinism or Humean scepticism (Knight 2004, 20–36). Michael Ruse has argued that design is a helpful metaphor in order to remind us about classic arguments for complexity (Ruse 2003). I am less convinced that design language is all that constructive, even if qualified in metaphorical language, for, while I agree with Ruse that notions of God as the divine "designer" directly intervening in the world are untenable, notions of design, even when used metaphorically, can conjure up unhelpful models such as that of "intelligent design" that he is also equally anxious to refute. In arguing for a recovery of natural law, I am not intending to argue that more "naturalistic" versions of natural law have a specific *requirement* to be linked with contemporary evolutionary theory. Rather, this article is asking a different, more modest question: namely, what might be possible ways of thinking theologically that are compatible with current ideas about evolutionary convergence? Are there avenues for finding some common ground with such ideas, without arguing for a fully fledged natural theology that has traditionally sought to argue for evidence for the existence of God through contemplation of the natural world? Traditionally, Christian theology has sought to distinguish between natural theology, which seeks to find God through reflection in the natural world, and revealed theology, which insists that God can only be known through divine revelation, more specifically understood as the revelation of God in the person of Christ. Theological reflection arises afresh in each new generation, even though it is deeply embedded in historical traditions. Theology may be developed from a faith perspective, as faith seeking understanding, but not inevitably so. And, given the eye of faith, what kinds of resonance might such evolutionary ideas have with theological considerations about divine providence and wisdom? I will be drawing particularly on Thomas Aquinas' theology in this chapter since he was a pioneer in his incorporation of a range of different areas of knowing in arriving at a synthetic approach to the truth about existence. While he used Aristotle's philosophy most extensively, he was also open to the most contemporary versions of scientific thinking available at the time and sought to respond to its truth claims in the light of the theological tradition that he had inherited.

Theology's roots in history is one of the most important distinguishing marks comparing theology with experimental science; the latter tends to assume that "old" science is simply the building blocks that may either be replaced by new discoveries or presupposed in building up a scientific picture of the world. Theology, on the other hand, deliberately trawls the tradition (and scripture) in order to reinterpret them in a fresh way; hence, it relies on a philosophy of interpretation or hermeneutics. Theology also incorporates scientific analysis of text into its hermeneutics, so it is perhaps more aware than evolutionary science of itself as a discipline that is culturally situated in a given historical context. The possibility of a link with evolutionary science is striking since evolutionary science, unlike many other experimental sciences, necessarily does concern itself with history, looking back to the dawn of existence in order to look ahead. There are also some things that can be said theologically that are more appropriate from within a particular faith tradition and frank acknowledgment of a starting point of faith. Hence, while there are points of "convergence" between evolutionary and theological thinking, there may also be glaring gaps, and these need to be recognized as such. Karl Rahner has suggested that theology and science inevitably disturb and threaten the other (Rahner 1983, 17). It is more likely in the present context that theology will be disturbed by evolutionary science, rather than the other way round! However, this will depend on how much and to what extent scientists are willing to acknowledge the possibility that there may be other sources of knowledge and truth claims that are outside the remit of experimental analysis. Rahner also suggested that theology can serve to confront those sciences that to make attempts to engulf other strands within science or, in this context, different disciplines or models over others. Theology provides a more modest claim to remind those engaged in dialogue of the relative weight of different truth claims. Theology, in this scenario, is one set of claims alongside others, rather than a "master" or "queen" of the sciences, pitched in judgment against it.

Contingency and "Laws" of Nature

Wolfhart Pannenberg in *Toward a Theology of Nature* argues strongly that a theology of nature needs to relate nature in its entirety to God, which includes a scientific understanding of natural processes (Pannenberg 1993, 73). Note that he used the term *theology of nature*, that is, a theological reflection on the significance of nature, unlike *natural theology*, an argument for God from contemplation of nature. Contemporary theologians, reluctant to "burn their fingers" have avoided dealing with the subject of nature. Pannenberg views any concept of ordered, fixed regularity of the universe as resting on mistaken Greek notions of the cosmos; instead, he suggests that the Hebrew notion of God is one that stresses the contingency of divine will. He asks if contingent occurrences in some sense also disclose regularity. The subject under discussion in this case was focused more on the "laws" of physics; but even these, he suggests, need to be considered under the category of contingency, for "only in this way would it be convincing that the order of the laws of nature on its part also is comprehended by the thought of creation and is not opposed to it" (Pannenberg 1993, 79). In other words, the historical experience of Israel is of a God who acts powerfully in the midst of contingent events, so that, while connections in occurrences arise, these only become visible from the end. Hence, he excludes the idea of purposefulness that directs everything *from the beginning*, in the sense of entelechy, for he believes that such forms of purposefulness amount to a loss of contingency. At the same time, he does allow for "partial development tendencies within the total process" (Pannenberg 1993, 83). Of course, his position means that, in some sense, the new enters *from ahead*, rather than from the past, so that forms are "overformed" by the new, rather than broken by them. This view is also implicit in his suggestion that the direction in evolution toward greater complexity comes through "field effects," rather than being implicit within the evolving species themselves. (Pannenberg 1993, 47). The concept of field effects seems to be taken from scientists who are on the more speculative end of the spectrum and who, in the biological sphere at least, are not very convincing. Pannenberg links field effects with spiritual energy, but this runs the danger of too close a marriage between a speculative scientific theory that is not well established and theology, a synthesis that he also criticises in the theories of Pierre Teilhard de Chardin. Pannenberg's theology of nature is successful inasmuch as it puts due emphasis on contingency in the natural world and could, thereby, be extended to include contingency in evolutionary processes. Such contingency is an accepted aspect of all evolutionary theory, so, in this sense, evolution is not a threat to theology as such. He has also managed to combine a theory about purpose with contingency, by directing purposefulness from ahead, rather than from the past. Yet it is worth asking if Pannenberg has been too ready to dismiss any understanding of directionality as implicit in the natural order, for his own rendering of purposefulness is necessarily transcendent, understood in eschatological terms, read into the history of nature in the light of experience. It is also worth asking if he has adequately considered the possible constraints within which evolution works, the subject of the present discussion. More particularly, we might ask if he has subsumed all understanding of general divine action of God into forms of special divine action. While the former makes more sense in the context of consideration of the natural world, the latter makes more sense in the context of human history. Pannenberg is no doubt reacting to the opposite more liberal tendency—that is, to deny any existence of special divine action.[1]

Laws of Nature and Natural Law

At this juncture, it is worth briefly drawing a distinction between the laws of nature and natural law and their respective meanings. Laws of nature may be interpreted as having a basis that is only partly described by the laws of science. At the outset, it is worth reiterating that biologists do not view "laws" of nature in the same manner as physicists, though the possibility of physical constraints within which evolutionary change takes place is also worth consideration. This is one factor in the kind of constraint that leads to convergence, but it is very unlikely to be the only factor. There are broadly four possible definitions concerning the physical laws of nature (Saunders 2002, 60–72). From an analysis of these laws, one definition is that they are a simple account of regularity or patterns, but this is not normally accepted since clearly identifiable regularities are not embedded in many physical laws. More instrumental accounts of laws of nature depict such laws as rational attempts to organize natural observed phenomena. Such an idea will lead to expectations of natural phenomena limited by the realm of possibilities open to the human mind. In other words, according to this view, the origin of the laws are not in nature itself in an ontological sense, but rather in the human mind. A third possibility is the necessitarian account of the laws of nature, which claims that physical laws ontologically determine which possibilities are open to the world and which are not. Observations achieved by science are reflections of this deep ontological structure of reality. Biologists tend to use the language of "necessity" in describing the patterning of the evolutionary process, though such a use of the term is somewhat careless, for it is very unlikely that they mean by this an ontologically structured "law" that pushes evolution in one direction rather than another. The final type of explanation is one that argues that laws of nature are irreducibly statistical in form; hence, the language of *probability* is one that fits most easily with scientific observations. Biologists have been far more reluctant to describe any of their observations in terms of "laws," mostly because the level of predictability is far less than that observed in physical science. One exception might be the Mendelian "laws" of genetics, but even these are subject to considerable variation and exception. Conway Morris' notion of convergence is more akin to the notion of directionality, rather than a "law" of nature; hence, it implies a measure of restriction within which evolutionary contingency operates, rather than resting on a physical or mathematical law as such. This is not to say that it operates outside the laws of mathematics and physics—this would be impossible—but rather that there is more to be said about convergence than simply a description in terms of physical laws.

I suggest that natural law, at least as devised in the classic tradition, is more consonant with evolutionary convergence and theological purpose compared with the laws of nature for a number of reasons. Natural law, in the first place, is related specifically to a goal or teleology in a way that laws of nature are not. While such a goal can be shorn of its theological origins, it makes sense in the context of the present discussion to consider such teleology from both a theological and a philosophical point of view. Second, natural law, at least since the late medieval period, acknowledges more specifically the element of interpretation by human beings, though, of course, the instrumental account of the laws of nature makes this claim as well. Third, natural law serves to set limits or boundaries for the activity of different forms of life, as will be explained further below. Natural law is not a fixed rule; hence, it has more in common with the way biologists understand "law" compared with physicists.

What Is Natural Law?

Natural law traditionally has been associated with ways of mapping the boundaries for human behavior, rather than a way into reflecting theologically about evolutionary theory. Natural law becomes in these formulations particular ways of interpreting human experience and action, rather than having its basis in ontological descriptions about the world. However, although contemporary forms of natural law have tended to isolate this concept and use it as a basis either for legal theory or ground a philosophical basis for ethics, the classic tradition rooted the theory very clearly in more general concepts about the intelligibility of the natural world. It is noteworthy that "new" natural law theory has attempted to sever natural law completely from its basis in the natural order (Biggar and Black 2002). In theological terms, natural law is also related specifically to the doctrine of creation. The grounding of natural law in the doctrine of creation means that, on the one hand, it has a clear ontological basis in the created order but, on the other hand, it also relies on the rational interpretive capacity of human beings. In this sense, it could be said to be situated midway between those philosophers such as Hilary Putman who argue for an ethics without ontology and more classical versions that argued that ethical frameworks have an ontological basis. It is not my intention here to enter into debate about which version of natural law theory is most appropriate for contemporary ethics, be it naturalistic or existentialist versions. Rather, natural law, like natural theology and natural science, derives from a realist account of the world, which makes more sense when understood in a way that is integral to the doctrine of creation, rather than split apart from it. Once this is appreciated, it becomes clear that reflection on natural law is suggestive of theological elements that are deep in the Christian tradition, rather than merely superfluous adjuncts in order to justify ethical mandates.

Natural law brings together three areas of human reasoning: namely, a study of biological nature as such, reason, and scripture. It is, therefore, highly suggestive of a way of mediating between theology and biological science in a manner that is not the case for more strictly philosophical concepts such as the laws of nature. The laws of nature are more useful in dialogue with cosmology or physics. Jean Porter suggests that natural law acknowledged the restrictions on human behavior as a result of biological limitations that pointed back to the earliest stage of human history prior to the formation of normative principles in conventions and customs. Hence, "these pre-conventional givens include the exigencies of our biological nature as well as reason, which is seen as setting both normative and practical restraints on human freedom, and Scripture, seen as a revelation of divine wisdom and will" (Porter 1999, 51). Such preconventional givens might include cultural traits that are found in species other than humans. Hence, convention seems to imply an advanced level of cultural consensus that is not found among nonhuman species, while the notion of preconventional givens, in the light of more recent research on whales or corvids, for example, might well include social patterns found in species other than our own. (Emery and Clayton 2004, 1903–7; Clayton and Emery 2008; Whitehead 2008). Yet, while natural law as grounded in nature implies a sense of restriction at one level, it also allows for a flexibility of interpretation at a secondary level of specific precepts; hence, it can be adapted to allow for new areas of understanding. This resonates specifically with the concept of convergence, where there is both a restriction toward the evolutionary appearance of certain forms but considerable flexibility in outcomes. In addition, natural law puts due emphasis on the continuity between animal and human behavior. This is not a form of "naturalism," in the sense of reading human behavior *out of* that found in animals, but rather an interpretation of human morality as a rational purposeful expression of tendencies found more generally in animals.

There was considerable variation in the way natural law was interpreted even in the middle ages (Porter 1999, 76–77). It was associated with that which was common to humanity and animals but also included, in some cases, the laws of the nations, the divine law in the prophets, Mosaic law, a human tendency to do good and avoid evil, and the concept of natural justice. It is important to emphasize that, even at this stage, the activities found in nature were considered to be reasonable; some even defended the idea in summary by suggesting that "nature is reason" (Porter 2005, 71). I will argue that, for the purposes of this discussion, the version of natural law defended by Thomas Aquinas is instructive since he articulated his understanding of natural law in the context both of an awareness of the philosophical and scientific debates of his time and theological reflection. While his understanding of biology was severely limited by the knowledge of the period, a fact that needs to be acknowledged, his interpretation of natural law still, I suggest, has elements within in it that are of significance. The medieval Scholastics were also much more prepared to admit the possibility of dispute than is sometimes presumed; hence, their views were far more open to the possibility of change than has often thought to have been the case. Public disputations allowed for the airing of all possible alternatives before arriving at a conclusion. Much of this is lost if we think of the medieval period as simply following rules and regulations, for the processes involved in arriving at these was far more contested than we might presuppose.

Aquinas believed that the purposeful behavior in nature was directed toward a good end, and such purposefulness was under the providence of God. His notion of primary and secondary causes allowed for a relative autonomy of the natural world, but it was one that was ultimately an expression of God's goodness as Creator. The first principle of natural law is "that good is to be sought and done, and evil to be avoided; all other commands of natural law are based on this" (Aquinas 1966: Qu. 94.2) The secondary principles of natural law include, first, that the natural tendencies of human beings correspond with that found in the "laws of nature," such as the tendency for self-preservation. Aquinas is referring here to all life forms, including plants. Second, there is a correspondence with that which "nature teaches all animals," including drives toward reproduction and rearing of young, while, third, there is a correspondence with that which is specific to rational animals, namely, an appetite for the good in rational terms. Note that there is nothing here to restrict rational animals to humans, though Aquinas had limited knowledge about this possibility in other species. He was prepared to suggest that, at times, humanity behaved in a way that was lower than "brute beasts," so that, although the hierarchy was intact, it had more fluid boundaries than we might imagine from his more negative mandates toward the treatment of animals. It is important to note both the distinctive characteristics that emerge at each level of complexity, as well as continuity. Of course, it is possible to redefine intelligence in such a way that plant life is included as well, as Anthony Trewavas has elegantly pointed out (Trewavas 2008). It seems that, in the latter case, the tendency for self-preservation is what this intelligence amounts to, a point also noted by the medieval scholars but not expressed in such terms, for they were not aware of the sophisticated means and communication pathways through which this could come about. The ability to distinguish self from another, even at this level of organization, is remarkable. One might prefer to name this decentralized intelligence "protointelligence," since, while it has many of the features, including mobile transport of information commonly associated with intelligent action, it is not developed to the same extent or to the same level compared with other species, such as mammals and birds. The main issue here is to note the way natural law is grounded in the biological structures of organization and behavior, right down to the simplest life forms.

Aquinas also was prepared to admit that natural law could be changed in its particular formulations, apart from the first principle that good is done and evil is avoided. This would imply, of course, that it is entirely permissible to update his understanding of biological processes and the secondary principles of natural law in the light of contemporary evolutionary science. He used the most recent scientific understanding available at the time; it is his *method* that is worth particular attention, for he allowed all aspects of the debate to be considered before arriving at his distinctive theological position. While this is closely related to natural theology, its intention is very different, for it is not simply about finding "evidence" in the natural world for God's existence but a way of enlarging an understanding of God's action in the world based on all forms of knowledge and in the light of revealed knowledge. It is a theology of nature, rather than a natural theology. He was also particularly insistent that, in the area of practical reason, there is room for contingency, so that "the more we get down to particular cases the more we can be mistaken" (Aquinas 1966, Qu. 94.4). In addition, human sinfulness means that, while natural law will direct human beings toward the good to some extent, its application to particular acts always falls short, which shows that Aquinas was no moral naturalist. (Aquinas 1970, Qu. 113.1)

It is also important to understand what Aquinas meant by the good being sought. He seems to mean the good as perceived by a particular creature, as it seems to them, expressed both in terms of "purpose" (intention) and intelligibility. All living creatures will try to pursue this good; even those who commit evil do so on the basis that it *seems* good as far as they are concerned. In this way, natural law does not suffer the same problems as ideas about design, for it might seem incongruent that one creature is designed for attack and one for defense. The pursuit of what seems good and the avoidance of evil are the meanings of the first principle of natural law. Of course, it might be possible to undertake a psychological study in order to analyze scientifically how far those who commit crimes actually believe that they are in some way benefiting themselves or not. Yet to try to do so is really to miss the point of natural law: the value of goodness is a philosophical good, even though more recently authors such as Martha Nussbaum (2001) have challenged its ontological status. Hence, to try and force goodness and evil into the category of experimental science is to make the same kind of category mistake as finding evidence for God from the "design" or workings of the natural world. Theology and evolutionary science can come together in some respects, but some questions are answerable only in one or the other category. Natural law provides a fragile bridge between the two areas, but the gap remains intact. Some, like natural lawyers, prefer to keep theological discussion out of natural law. Others wish to sever natural law's links with biology as suffering from too close an affiliation with "naturalism." The naturalistic fallacy long despised by philosophers ever since David Hume and G. E. Moore looms large: that is, the assumption that what is is also automatically good.

Yet there are three points to be made in response to this. The first is that the naturalistic fallacy is based on the premise that there can be a clear separation between descriptive accounts of being and evaluation, which itself presupposes a dualistic separation between subject and object. Philippa Foot has been arguing against the reasoning that splits facts and values presupposed in the naturalistic fallacy for a number of years (Foot 2001; 2002, 1–2). Hence, the acceptance of the naturalistic fallacy presupposes forms of dualism that follow from Enlightenment philosophy. The second is that the whole of the created order is not endorsed as having moral goodness in the manner anticipated by stronger versions of naturalism. For while Aquinas understood the created order to be both intelligible and good, there were always elements of contingency and fallibility. Third, if we look more carefully at Aquinas' interpretation, it is clear that the goodness of natural existence is derived not simply from biology as such but from the relationship between God, understood as transcendent, and creature. Hence, in order to discover the way natural law was treated in the classic tradition, we need to ask what is the manner in which natural law can become directed toward true goodness, which for Aquinas is a theological goal. This question can only be answered in relation to Eternal law, for natural law is defined as participation in Eternal law by rational creatures.

Eternal Law and the Wisdom of God

Aquinas allows for all life forms to share in Eternal Law or Reason through participation. It is here that Aquinas (and the classic tradition generally) parts company with those forms of evolutionary philosophy that view God as somehow emergent *from* the evolutionary process. However, rational creatures are able to share in Eternal Reason in a reasonable and intelligent way, which is impossible for non-rational creatures (Aquinas 1966: Qu. 91.1). Aquinas also associates the Eternal Law with Divine Wisdom, so that:

through his wisdom God is the founder of the universe of things, and we have said that in relation to them he is like an artist with regard to the things that he makes. We have also said that he is the governor of all acts and motions to be found in each and every creature. And so, as being the principle through which the universe is created, divine wisdom means art, or exemplar, or idea, and likewise it also means law, as moving all things to their due ends. Accordingly the Eternal Law is nothing other than the exemplar of divine wisdom as directing the motions and acts of everything.

Aquinas 1966: 93.1

Of course, such an understanding might seem to deny the possibility of contingency in the natural order, though Aquinas resists such a suggestion by his formulation of God's acting in an *analogous* way to human laws' acting in human hearts, so that "God impresses on the whole of nature the principles of the proper activities of things . . . ," and "the impression of an inward active principle is to the things of nature what the promulgation of law is to men, for by this, as we have argued, a certain directive principle is imprinted on human acts" (Aquinas 1966: Qu. 93.5). Where natural processes seem to "fail," this is the result of "interruptions" to ordered patterns of particular causes rather than interruptions in universal causes. His notion of eternal law as the guiding principle under which all other laws are subsumed ensures that his understanding of natural law is both grounded in nature and, at the same time, thoroughly theistic.

Natural Wisdom and Natural Law

For Aquinas, wisdom is more specifically a virtue that can be acquired as well as a gift of the Holy Spirit: true wisdom is knowledge of an ultimate good end, while false wisdom is fixed on material goods (Aquinas 1972: Qu. 45.1). The first stage of wisdom is to shun evil, while its last stage is to bring all things back to its rightful order, under acts of charity (Aquinas 1972, Qu. 45.6). Of course, the order or chain of being in which Aquinas situated the place of humanity needs to be adjusted in the light of evolutionary knowledge and the more qualified role that humanity now has in the overall evolutionary process. This is where theological models that identify evolution with "progress" fall short. Michael Ruse has criticized Holmes Rolston for equating evolution with "progress," though such models are not inevitably anthropocentric (Ruse 2003, 308–11). The gradual increase in evolutionary complexity should not be equated with notions of "progress," since this implies a linear direction for evolution that is scarcely tenable. Yet there is nothing intrinsic in Aquinas' position that would prevent such adjustment. Even his understanding of eternal law, while perhaps suggestive of more "Platonic" notions of form, could be viewed in more probabilistic ways that put much more emphasis on the place and importance of contingency. Aquinas also distinguishes between the gift of wisdom, which operates in matters of faith, and the virtue of wisdom, which acts in matters of grasping first principles of thought. Both forms of wisdom are about rightness in judging according to divine norms. As a gift, wisdom arises from charity that unites the believer with God. Inasmuch as natural law represents participation in the eternal law, or divine wisdom, so the virtue of wisdom and the gift of wisdom facilitate the movement toward the good purpose implied though natural law. Aquinas restricts his discussion of wisdom to rational beings, for both the speculative virtue of wisdom and its practical counterpart in prudence (practical wisdom) are intellectual virtues. However, just as the eternal law can be said in a manner of speaking to be "imprinted" in some sense on the whole of the natural order, so too a form of natural wisdom could be said to exist through his notion of participation of all creatures in the divine reason, or wisdom, for "non-rational creatures . . . participate in the divine reason by way of obedience: the power of divine reason extends to more things than comes under human reason" (Aquinas 1966: Qu. 93.5). Indeed, one could say that all creatures are thereby given an imprint of the Trinity, so that "in all creatures, however, we find a likeness of the Trinity by way of trace in that there is something in all of them that has to be taken back to the Divine Person as its cause" (Aquinas 1967: Qu. 45.7). The use of terms such as *law* or *wisdom* in nonhuman creatures in Aquinas becomes figurative for "non-rational creatures do not hold law as perceiving its meaning, and therefore we do not refer to them as keeping the law except by figure of speech" (Aquinas 1966: Qu. 91.2).

Yet it is also clear that, for Aquinas, wisdom represents a higher level of perception than that possible through synderesis or our natural reasoning processes. This applies more specifically to ethical action, so that judgments may be immediately obvious to human reason, while other judgments are the result of more careful consideration of the wise: "these indeed, belong to the law of nature, but as necessitating instruction on the part of ordinary people by the wise. . . . Lastly there are actions to judge of which human reason needs divine instruction, which teaches us about the things of God" (Aquinas 1969: Qu. 100.1). This hierarchy of thinking insisted that, while natural reasoning can take us a certain distance, "what belongs to faith is above natural reason" (Aquinas 1969: Qu. 100.1). This is an important strand in his *Summa*, for it shows that natural law understood as participation in the Eternal Law only really makes sense from the perspective of faith. While elements may be obvious to the common reasoning of all rational creatures, including wisdom as learned, wisdom as gift is only possible from the perspective of faith, for "the gift of wisdom presupposes faith, since a man judges well what he already knows. . . . Piety is wisdom and for the same reason also is fear. If a man fears and worships God he shows he has a right judgement about divine things" (Aquinas, 1972: Qu. 45.1).

Some Tentative Conclusions and Questions

I began this discussion with consideration of Pannenberg's emphasis on God as contingent, where the laws of nature become known in retrospect, superimposed, as it were, from the known future that is in God. His view fits more closely with the probabilistic understanding of the laws of nature that is compatible with current natural, physical science. While debates in evolutionary theory are not sufficiently clear to be delineated into alternative laws, the schema is helpful in that viewing convergence in probabilistic terms makes more sense compared with patterning, deterministic, or instrumental alternatives. The idea of patterning is too suggestive of the concept of fixity of design in the natural order, which is unhelpful as it is suggestive of a fixed cosmos. Of course, there are elements in the thought of Aquinas that do point to rather too great a fixity in the ordering of nature that many would feel uncomfortable with today. The strangeness of the medieval period with its very different understanding of cosmology needs to be acknowledged. However, any such accusations of "Platonism" need to be tempered by the realization that both Aquinas' belief in secondary causes and his notion of eternal law as analogous to human law are more suggestive of a framework within which contingency can move, rather than anything more rigid. The question now becomes whether Pannenberg over reacted against the possibility of God's working through an ordering process implicit at the beginning of creation. It is here that I have suggested that the concept of natural law is helpful, both because its meaning has been compressed and reduced in contemporary discussion, severed from the doctrine of creation, and also because it provides a way of thinking positively about biological processes from a theological point of view.

The natural law in all creatures was associated with a purpose or telos toward the good end in God. Aquinas' understanding of the created order was in terms of a hierarchical chain of being. Such an understanding clearly needs to be challenged in the light of evolutionary theory. However, his concept of natural law links all processes of life with human life in a way that affirms the connectivity of all life forms and, more specifically, with the life possible through participation in God. He describes the work of the wisdom of God in the natural order in terms of an artist, bearing traces of the Trinity in its unfolding. While his biological understanding was outdated, his perception of God as one who makes impressions on the natural world is still compatible with evolutionary contingency. More important, perhaps, his notion of natural law—and by implication, natural wisdom—provides a theological interpretation of the possibility of convergent forms and evolutionary "purpose." In addition, it is important to note that his understanding of the created order allows for its unfolding without an imposition of quasidivine intervention; the sense in which God governs is through a bestowal of inherent properties that direct creatures toward a given end.

There are, of course, a number of questions left unanswered that need to be addressed. These include the question of how far Aquinas' notion of seminal forms is compatible with evolutionary contingency. In this, I suggest that he had a biologically naïve view, but, given that he was writing in the twelfth century, this needs to be taken into account. His intention was to allow his theological reflection to stand up to scrutiny in the light of current biological knowledge, which, of course, in those days was intricately linked with philosophical reflection. It is also important to distinguish concepts from changeable realities. For example, the emergence of human beings on the evolutionary tree does not mean that human nature as such does not exist, or any other species for that matter, rather that each species has the potential to either change into something else or become extinct. Hence, it is perfectly reasonable to accept that given characteristics for living species exist while arguing that biologically such forms may have derived from other forms or may even disappear in the future. Aquinas was not aware of the possibilities of extinction of forms, but that does not mean that the concept of species as such is now totally redundant. The existence of convergence and parallelisms show up significant similarities between evolving species, but their differences and individual characteristics ought not to be forgotten.

Is Aquinas' understanding of wisdom too anthropocentric from the perspective of current biblical knowledge? The book of Proverbs invites its readers to "Go to the ant, you sluggard, See its ways and be wise" (Prov. 6:7). The seeing is not so much detailed observation of information about the ants, implied perhaps in Nigel Franks' account of the workings of an ant colony (Franks 2008) but, as Norman Habel suggests, *perceiving* the inner distinctive core of what it is to be an ant. (Habel 2003, 281–98) In other places in wisdom literature, the phrase *to discern* (*bin*) is used, often following the act of seeing, to describe the *process* of becoming wise. Hence, discernment is integral to what it means to gain wisdom. Discernment considers a range of options but ultimately lights on "the way," understood not just as the alternative between two paths but also as the inner "driving" characteristic of something. It is noteworthy that, for a number of contributors to this volume, one had the impression that the biologists concerned were able to imagine actively what it might be like to be the organisms that they were studying, to perform, as Barbara McClintock suggested in her long-standing relationship with maize plants, a "feeling for the organism" (Keller 1983). Significantly, in the book of Job 28, the characteristic of finding wisdom *also* applies to God, as God "sees" the different components of creation. This suggests a degree of freedom to creatures and a form of natural wisdom that goes even further than that implied by the more "top down" approach of participation in Eternal Wisdom that Aquinas suggests. Hence, both contemporary biblical studies and biology suggest that rather more emphasis needs to be placed on the concept of organisms as separate selves, with their own degree of "wisdom." Yet, I would hesitate to take this as far as Alfred North Whitehead does in his process philosophy, for his suggestion of a "mental pole" in all existence, even in the very fabric of material reality, does not connect readily with the common experience among biologists (myself included here) as to the crucial difference between life and nonlife. The advantage of Aquinas' understanding of natural law is that it *does* make this distinction, even while acknowledging that all of creation, including the material world, is under the providence of God.

Is Aquinas' belief that all of creation naturally orientates itself to the good too idealistic? Or does it imply that purposefulness under divine providence amounts to progress? This is the philosophical alternative with its theological counterpart to the argument from design suggested by Michael Ruse (Ruse 2008). It is, nonetheless, incorrect to identify Aquinas with arguments from design, for his view of the relationship between God and nature was very different from that of William Paley. Aquinas also had little intention, unlike many of his commentators imply, of providing secure "proof" of God's existence from the natural world, in spite of his infamous "Five Proofs," that were written specifically with unbelievers ("Gentiles") in mind. Jean Porter acknowledges this difference by suggesting that

Paley's argument turns on an analogy between artefacts and living creatures, whereas Aquinas, like Aristotle before him, insisted on a contrast between them. On their view, artefacts must be designed and assembled precisely because they do not possess their own intrinsic forms, or correlatively, their own internal orientation towards a purpose intrinsic to that form. . . . Aquinas' specific argument . . . does not appeal to the design of living creatures, but rather to the goal-directed character of natural operations, including, but not limited to the operations of living things.

Porter 2005, 87

Rather, Aquinas wanted to take human reasoning as far as it could go toward God, but, in the end, he recognized clearly that a step of faith was always required in order to understand life in God. The purposefulness in the created order could, as far as Aquinas was concerned, have come from the realization of its own nature, inherent within it. In the human sphere, nature and grace were interrelated: grace builds on nature, rather than denies it, as Rahner has expressed so clearly in his more contemporary theological analysis. Aquinas was also not writing out of the context of Enlightenment doubt, which was to follow much later in history. He is, however, optimistic in his account of the possibilities for human nature and of the value of all creatures more generally. The goodness in creatures was manifested to the extent to which it demonstrated purpose and intelligibility. Yet he insisted that the goodness in human nature could only be discovered through the grace of God and that such theological knowledge would always be outside human grasp because of human sinfulness. I suggest that his view that sinfulness makes evil seem good is realistic rather than utopian. Even the perpetrators of the worst atrocities known to humankind have normally worked under the illusion that they are doing something that is a good, either for themselves or for the causes to which they are committed. He did not, however, have the knowledge of either psychology or evolution to recognize adequately the full extent of what one might term "natural evil." Clearly, such an account needs to be brought into an understanding of theodicy,[2] but it is not inherently more challenging than similar accounts that have to take on board humanity's inhumanity to itself and other creatures.

Is the possibility of a recovery of natural law as linked with the natural world and God possible in a postmodern context with its emphasis on the deconstruction of any "essentialist" notions of either God or nature? Given the echoes of Plato in Aquinas' notion of eternal law and wisdom in God, how far is this compatible with contemporary understanding about God? What might be the meaning of the providence of God in this context? I suggest that, inasmuch as biologists work from the presumption that what they are discovering does have some basis in ontological reality, so too a theology of natural law reaffirms this assumption. In this, natural law and biology are on common ground. However, both theologians and biologists do have to take into account the historical contingency of their work as being limited by context and situation. Having said this, there is no *a priori* reason to exclude as a matter of course those attempts to put tentative theoretical bones onto the debate—that is to grope toward a theory that encompasses the natural world. While we need to be modest, perhaps, about the role of the human mind in such constructions, to assume that there is no contact with reality is equally presumptuous, as it makes the assumption of radical relativism. Perhaps the most we can be content with is to discuss the issue in terms of probabilities rather than certainties.

In this, the providence of God toward goodness is one that is accepted as probable on the basis of faith, always tinged with doubt, rather than "proved" through reflection on the emergence of evolutionary complexity. Such providence cannot be identified with human progress either, for such a presumption assumes (falsely) that we know the mind of God in its entirety.

Is it permissible to recover medieval concepts isolated from their original context and concerns? What is the relation between natural law and natural theology? Given the arguments for a rerooting of natural law in a doctrine of creation and its resonance with theories of evolutionary purpose, what are the implications for debates about the relationship between evolutionary theory and moral agency? It is clearly not feasible to lift Aquinas' teaching from its original context without some adjustment to contemporary beliefs and practices. It is also important not to come to too hasty an accommodation with contemporary beliefs. However, I suggest that, while these questions do need to be addressed, they are not ultimately destructive of the thesis presented here, namely, that the concept of natural law provides one way of understanding in theological terms what evolutionary science is hinting at through notions of convergence and evolutionary "purpose." It also offers considerable advantages over eighteenth-century alternatives that viewed the order in creation as analogous to a watch made by God or to comparable forms of natural theology that identified too readily physical processes in nature with divine purpose and intention. Aquinas was always modest about the possibility of knowing fully the eternal reason in God: such knowledge could not be attained in this life, so that only God and the blessed can know the Eternal Law, but all rational creatures can see its effects, understood in terms of natural law. In this, it is vital to keep the *apophatic* tradition alive, as well as the *cataphatic* tradition.[3] Aquinas admitted that, toward the end of his life, he believed that all his previous intellectual work was "as straw" in the light of more mystical experiences of God that he had only dimly glimpsed. This is not to deny the importance of theological reflection but rather to qualify its place. In the end, our understanding and reasoning can only take us so far; this is inherent in Aquinas' *Summa*. However, the area of unknowing in God is ungraspable: it certainly cannot be grasped through a study of biological reality. God is always not so much something that can be arrived at through science but an existential Who that challenges those who seek to find such an encounter. In this, theology—meaning literally, language about God—makes small steps to understand more about God, but, like biology, its area of knowing is always incomplete. Unlike biologists, who hope perhaps that one day it will be possible to know all there is to know about the natural world, theologians, true to their task, would do well to be far more modest, for in the light of the infinity of God, finite human attempts to grapple with such reality may seem paltry indeed. Perhaps the wonder of so much that is not yet known in the biological world can be seen in one sense as an implicit religious experience, reflecting that which biology has itself recommended as necessary for human survival. In this, theology and biological science may converge from very different origins, for wonder, like intelligence, is integral to what it means to be human.

Acknowledgments

I would like to thank Simon Conway Morris for the initial invitation to take part in this project and for helpful feedback, Mary Ann Meyers for her part in the project's coordination, the John Templeton Foundation for its financial support, and George Coyne of the Vatican Observatory for his hospitality. I am also grateful to Jean Porter for allowing me to read parts of her book while still in press. This chapter was previously published in *Zygon* 42 (December 2007): 981–98.

Notes

1. For a discussion of the distinction between special divine action and general divine action, see Saunders 2002, 18–32.

2. Theodicy is the theological attempt to justify how to reconcile the belief in the goodness of God with the presence of evil of the world.

3. The *apophatic* tradition claims that we know God by stating what cannot be known of God, while the *cataphatic* tradition is more positive about its claims about what can be known of God, such as God is goodness and love.

References

Aquinas, T. 1966. *Law and political theory*. Trans. Thomas Gilby. *Summa Theologiae*, vol. 28, 1a2ae. London: New Blackfriars.

———. 1967. *Creation, variety and evil*. Trans. T. Gilby. *Summa Theologiae*, vol. 8, 1a. London: Blackfriars.

———. 1969. *The old law*. Trans. David Burke and Arthur Littledale. *Summa Theologiae*, vol. 29,1a2ae. London: Blackfriars.

———. 1970. *The world order*. Trans. M. J. Charlesworth. *Summa Theologiae*, vol. 15, 1a. London: New Blackfriars.

———. 1972. *The consequences of charity*. Trans. Thomas Heath. *Summa Theologiae*, vol. 35, 2a2ae. London: Blackfriars.

Biggar, N., and R. Black, ed. 2002. *The revival of natural law: Philosophical, theological and ethical responses to the Finnes-Grisez school*. Aldershot, UK: Ashgate.

Clayton, Nicola S., and Nathan J. Emery. 2008. Canny corvids and political primates: A case for convergent evolution in intelligence. In *The deep structure of biology*, ed. S. Conway Morris, 128–42. West Conshohocken, PA: Templeton Foundation Press.

Conway Morris, Simon. 2003. *Life's solution: Inevitable humans in a lonely universe*. Cambridge: Cambridge University Press.

———. 2008. Evolution and convergence: Some wider considerations. In *The deep structure of biology*, ed. S. Conway Morris, 46–67. West Conshohocken, PA: Templeton Foundation Press.

Emery, N. J., and N. S. Clayton. 2004. The mentality of crows: Convergent evolution of intelligence in corvids and apes. *Science* 306: 1903–7.

Foley, Robert. 2008. The Illusion of purpose in evolution: A human evolutionary perspective. In *The deep structure of biology*, ed. S. Conway Morris, 161–77. West Conshohocken, PA: Templeton Foundation Press,

Foot, P. 2001. *Natural goodness*. Oxford: Oxford University Press.

———. 2002. *Moral dilemmas and other topics in philosophy*. Oxford: Clarendon Press.

Franks, Nigel. 2008. Convergent evolution, serendipity, and intelligence for the simple minded. In *The deep structure of biology*, ed. S. Conway Morris, 111–27. West Conshohocken, PA: Templeton Foundation Press.

Habel, Norman. 2003. The implications of God discovering wisdom in Earth. *In Job 28: Cognition in context*, ed. Ellen van Wolde. Boston: Brill Leiden.

Keller, Evelyn Fox. 1983. *A feeling for the organism: The life and works of Barbara McClintock*. New York: Freeman.

Knight, David. 2004. *Science and spirituality: The volatile connection*. London: Routledge.

Nussbaum, Martha. 2001. *The fragility of goodness*. 2nd ed. Cambridge: Cambridge University Press.

Pannenberg, Wolfhart. 1993. *Toward a theology of nature: Essays on science and faith*. Louisville, KY: John Knox Press.

Porter, J. 1999. *Natural and divine law: Reclaiming the tradition for Christian ethics*. Grand Rapids, MI: W. B. Eerdmans.

———. 2005. *Nature as reason: A Thomistic theory of the natural law*. Grand Rapids, MI/ Cambridge: Eerdmans.

Rahner, Karl. 1983. *Theological investigations*. Trans. E. Quinn. Faith and Ministry, vol. 19. London: DLT.

Ruse, Michael. 2003. *Darwin and design: Does evolution have a purpose?* London: Harvard University Press.

———. 2008. Purpose in a Darwinian world. In *The deep structure of biology*, ed. S. Conway Morris, 178–94. West Conshohocken, PA: Templeton Foundation Press.

Saunders, N. 2002. *Divine action and modern science*. Cambridge: Cambridge University Press.

Trewavas, A. 2008. Aspects of plant intelligence: Convergence and evolution. In *The deep structure of biology*, ed. S. Conway Morris, 68–110. West Conshohocken, PA: Templeton Foundation Press.

Whitehead, Hal. 2008. *Social and cultural evolution in the ocean: Convergences and contrasts with terrestrial systems*. In *The deep structure of biology*, ed. S. Conway Morris, 143–60. West Conshohocken, PA: Templeton Foundation Press.

12 PURPOSE IN NATURE

On the Possibility of a Theology of Evolution

John F. Haught

The possibility, ever so distant, of banishing from nature its seeming purpose, and putting a blind necessity everywhere in the place of final causes, appears . . . as one of the greatest advances in the world of thought. . . . To have somewhat eased the torture of the intellect which ponders over the world-problem will, as long as philosophical naturalists exist, be Charles Darwin's greatest title to glory.

E. Du Bois-Reymond 1876, cited in Towers 1969, 78

It has been said that all scientists have a secret passion for teleology but that, like a mistress, she has to be kept out of sight of polite company. For myself, I would be happy to take her into public as a respectable married woman, provided that I am allowed to specify in what sense I am using the term.

Bernard Towers 1969, 88

At issue in most encounters of science with religion are two large questions: (1) is nature all there is; and (2) does nature have a purpose? How you answer the first question, of course, will determine whether you will even bother with the second. For, if nature is all there is, there can be no point in asking whether it has any point to it. Nature would have neither first nor final cause. Life would evolve, but no intentionality or overarching meaning would accompany its unfolding across time. *In the final analysis,* then, the whole of life and evolution would be a futile foray into the void.

If you answer "Yes" to the first question and "No" to the second, you are a follower of the philosophy or belief system known as *naturalism.* And if you think science supports the two propositions, you are a "scientific naturalist." When I speak of naturalism in this essay, I mean "scientific naturalism." This is a common belief in the academic world, so much so that the Duke University philosopher Owen Flanagan can claim that the purpose of his discipline these days is to make the world safe for naturalism (Flanagan 2002).

Religions and theologies, of course, reject naturalism. To be more precise, for my purposes here, it can be said that at least *theistic* religions such as Judaism, Christianity, and Islam definitively repudiate naturalism. For them, nature is certainly not all that exists, nor can the cosmos be without purpose. And yet, where is there any clear evidence of purpose? This is what the naturalist wants to know. Moreover, if cosmic purpose were to manifest itself vividly anywhere in nature, wouldn't it be in the life world? Yet it is here especially that science today seems to provide anything but support for the religious sense that purpose exists in nature. In fact, Darwinian biology finds at best only an *apparent* purposiveness in the adaptive living design that reportedly came about blindly and unintended (Ruse 2003). Conventional evolutionary accounts of life seem only to confirm the modern suspicion that the cosmos is pointless and impersonal—all the way down. It is true, the naturalist often admits, that humans have "purpose on the brain" (Dawkins 1995, 96), but nowadays the habit of looking for purpose can be accounted for apparently in Darwinian terms. The human inclination to believe that there is purpose in the world can be said to be an evolutionary adaptation (Wilson 1998, 262; Burkert 1996, 20), a crafty invention by our genes that, like all genes, are in the business of striving for immortality. Or, if not a direct invention of genes, the human intuition of purpose may be a parasitic complex feeding on brains designed for more practical purposes during the Pleistocene (Atran 2002, 78–79; Boyer 2001, 145). In either case, from the evolutionary naturalist's point of view, the religious habit of "projecting" meaning into the universe is *ultimately* explainable in biological terms *rather than* by the supposition that any divine presence actually exists and influences events in nature.

But what would the evidence for purpose look like if it did indeed exist? Could any conceivable "final cause" (Aristotle's name for purpose) ever show up in the domain in which science undertakes its distinctive kind of inquiry? Can scientific method alone ever decide the question of cosmic purpose? And what does the term *purpose* really mean, for example, when naturalists deny that the universe has a purpose to it? Just as the atheist cannot meaningfully deny the existence of God without having some idea of what *God* means, likewise naturalists cannot reject purpose without having some preconception of what they are ruling out.

Naturalists, as I read them, usually conflate the notion of purpose with that of "intelligent" design (e.g., Dawkins). Then, if they think they can account for the obviously complex design in living organisms naturalistically, as Darwinism seems to allow, they tell us that organisms only *appear* to be designed intelligently. Organic adaptations are "design-like," but they are not the outcome of any underlying purposiveness (Dawkins 1995; Ruse 2003, 268–70, 325). The appearance of intelligent design has come about as the result of a completely unintelligent Darwinian process consisting of three main ingredients: random genetic variations (now known as mutations), impersonal and blind natural selection, and many millions of years of time. As far as evolutionary naturalism is concerned, there is nothing even remotely intelligent in the causal background of adaptive living design. Moreover, even our own intelligence is the product of a completely unintelligent process (Flanagan 2002, 11).

However, from a religious or theological point of view, purpose is not the same thing as design. Design is too shallow an idea to express all that is implied in the idea of meaning or purpose. In this essay, I take purpose to mean not design but an *overall aim toward the actualizing of value*. What makes any process purposive is that it is bringing about an end or goal that is self-evidently worthwhile or *good*. For example, I take my writing a chapter for this anthology to be purposive since its intended goal is to accomplish something that I consider worthwhile. I am not alone, of course. Others who are contributing to this book project also consider their efforts to be purposive since they take the sharing of insights with fellow inquirers to be intrinsically good. Likewise, evolutionary naturalists like Richard Dawkins or Daniel Dennett think their own writings and careers are quite purposeful inasmuch as they believe their work is fostering the undeniable value of truth. If they did not take the dissemination of truth to be a goal worth pursuing, they would hardly care whether their readers took them seriously or not. Nor would they expend so much effort trying to persuade us that the religious sense of purpose is pure illusion. Clearly, then, science-minded naturalists also have purpose-on-the-brain. If they were to disagree with what I have just said, they would be sabotaging their clear intention of trying to make the world a more enlightened place. For naturalists, doing science is eminently purposeful since it promotes the self-evident value of truth.

Most naturalists, upon reflection at least, will agree that they, like other humans, are purpose-driven. They will not want to deny that human existence can be made meaningful when dedicated to something of undeniable value, such as the pursuit of truth. But is this admission enough to convince the naturalist that the universe itself is purposeful? And what about life in the universe prior to the evolutionary emergence of beings with "purpose-on-the-brain"? Was purpose operative in the natural history that produced so many diverse organisms and species in the biosphere during the billions of years before conscious organisms appeared? In order to begin a response to these questions, I believe we must expand our vision from one that focuses narrowly on biological evolution and ask about the wider *universe* that sponsors the Darwinian process. Life and evolution, after all, are features of a much larger cosmos than the rather restricted territory that evolutionary naturalists typically take into account. The question of purpose in biological evolution that this book is taking as its unifying theme must not be separated from the deeper issue of whether purpose can, in some sense, be attributed to the universe as a whole.

But is there purpose in the wider universe? Not surprisingly, the scientific naturalist responds that there is none and that Darwinism is definitive proof of it. Of course, as I have just shown, the naturalist will not deny that an individual's life may still be filled with meaning (see Flanagan 2002, for example). In fact, some naturalists go so far as to claim that the absence of purpose in the universe at large is what allows human lives to become all the more filled with purpose (Klemke 1999, 186-97; Gould 1977, 12–13). If people can learn to swallow the bitter proposition that the cosmos is devoid of inherent purposiveness, they can then begin to understand that whatever values and meanings exist must have their origins in our human creativity, and this should fill us with pride. As Owen Flanagan puts it, "It seems like good news that meaning and purpose are generated and enjoyed by me and the members of my species and tribe, rather than imposed by an inexplicable and undefinable alien being" (Flanagan 2002, 12).

I have not the space here to dwell on the dubious consequences, especially ecological, of this exceedingly anthropocentric perspective. Instead, I will ask about the credibility of the modern claim that Darwinian evolutionary biology rules out not only evolutionary but cosmic purpose as well. Since most evolutionists believe they can explain living design in a purely naturalistic way, it follows for them that any religious belief that the cosmos is here for a reason and that it is in the business of achieving something of undeniable value is an illusory projection by "purpose-on-the-brain" organisms unable to resign themselves to the universe's ultimate indifference. According to many naturalists today, the Darwinian formula for evolutionary diversity—a recipe that consists especially of blind chance, impersonal selection, and an enormous expanse of cosmic time—is enough by itself to account for the nonmiraculous emergence of complex living design, including human intelligence. There is no need to look for evidence of divine influence either in life or the cosmos in its entirety. The three central ingredients in the evolutionary recipe are incompatible with a religious trust in the meaningfulness of the universe.

Naturalists, then, can rightly expect from theology a credible account of why the universe as a whole would sponsor a life story in which there is such an abundance of contingency (undirected events that we refer to as accidents) along with impersonal selection and a seemingly "wasteful" amount of time, a set of features that contemporary biology takes as the essential setting for evolution. In view of the apparent success of Darwinian explanations of life, evolutionary naturalists now view final causal or teleological explanations as having been decisively routed (Rose 1998, 211; Cziko 1995). But is the apparent absence of teleology from nature something that science can ever claim to have discovered, or is it not rather the inevitable result of a naturalistic ideal of knowing that decides from the outset that purpose simply cannot fall within the realm of knowable being?

The rather puritanical naturalist outlook, after all, is even annoyed with common sense, and not just with religion, because of our irrepressible tendency to ask about the meaning or purpose of things. Listen, for example, to these words of the renowned Harvard professor and evolutionary naturalist Richard Lewontin:

> Our willingness to accept scientific claims that are against common sense is the key to an understanding of the real struggle between science and the supernatural. We take the side of science . . . because we have a prior commitment, a commitment to materialism. It is not that the methods and institutions of science somehow compel us to accept a material explanation of the phenomenal world, but, on the contrary, that we are forced by our *a priori* adherence to material causes to create an apparatus of investigation and a set of concepts that produce material explanations, no matter how counterintuitive, no matter how mystifying to the uninitiated. Moreover, that materialism is absolute, for we cannot allow a Divine Foot in the door.
>
> Lewontin 1997, 31

To Lewontin and other naturalists, science is unintelligible apart from an *a priori* philosophical commitment to materialism. Lewontin concedes, however, that his commitment to materialism is not itself a conclusion based on scientific discovery but a sheer profession of faith. So we may also agree that it is not necessarily science per se but *materialist ideology* that is irreconcilable with theological (and that also means teleological) explanation.

It is entirely conceivable, then, that the cosmos as seen through the belief system known as naturalism (which is more often than not materialist in Lewontin's sense) has left something real out of its picture of reality. On careful inspection, naturalism itself turns out to be a construct erected on an avowedly *faith-filled* decision that only a method of inquiry that leaves out any reference to purpose is intellectually acceptable.

A Wider Way of Seeing

Perhaps, though, there are wider ways of seeing and understanding than the one Lewontin proposes as normative. Maybe there is even a richer empiricism than the restrictive kind operative in conventional science. I shall propose here that conventional scientific empiricism fails to notice even the obvious fact of our own intelligent subjectivity and this is a major reason why it cannot "see" purpose either in evolution or the universe.

To understand the *real* world, I suggest, we need to attend carefully not only to the objectifiable world but also to the reality of our own inner experience. After all, our own experience and consciousness are not alien to nature. They are nature's "inner side," as Teilhard de Chardin emphasizes. Lewontin's commonly accepted version of naturalism anxiously suppresses our common-sense awareness that our minds are indeed real, and it tries to place the entire sphere of subjective experience beyond the pale of truthful knowing. Each one of us, including the scientist, already knows that subjective experience, our own as well as that of many other living beings, is a fact of nature. But, as Teilhard correctly observes, naturalism's restricting of conscious subjectivity "to humans and perhaps a few other forms of life has only served as a pretext for eliminating subjectivity from [science's] *general* picture of the universe." Naturalists have looked upon consciousness as a "bizarre exception, an aberrant function." Immediate evidence of consciousness, Teilhard observes, appears only in the human domain. But, for some unjustifiable reason, scientific naturalism has assumed that our subjectivity is an "isolated case" and, hence, "of no interest to science" (Teilhard 1999, 23–24).

Now, however, as Teilhard argues, we must enlarge our vision:

Evidence of consciousness appears in the human, we must begin again, correcting ourselves, therefore half-seen in this single flash of light, it has cosmic extension and as such takes on an aura of indefinite spatial and temporal prolongations.

Indisputably, deep within ourselves, through a rent or a tear, an interior appears at the heart of beings. This is enough to establish the existence of this interior in some degree or other everywhere forever in nature. Since the stuff of the universe has an internal face at one point in itself, its structure is necessarily *bifacial*; that is, in every region of time and space, as well, for example, as being granular, *coextensive with its outside, everything has an inside.*

Teilhard 1999, 24

One way of evading Teilhard's challenge to make subjectivity part of our general world picture is to become a dualist. The dualist, following Descartes, does not deny the reality of mind or subjective experience but separates subjectivity altogether from the objective universe. Naturalists such as Lewontin, however, cannot take this route since they view such dualism as a subterfuge for a supernaturalism antithetical to their creed. There can be no arena of consciousness apart from the rest of nature since, for naturalism, nothing other than nature actually exists. Naturalists, therefore, are in the peculiar position of knowing that their own subjective experience is performatively real, and yet they cannot find any space for it in nature. It is my view that the naturalist expulsion of purpose from nature is of a piece with its refusal to let subjectivity, including our own intelligence, be seen as part of the real world.

Of course, scientists as such are perfectly justified in excluding *methodologically* any considerations of subjectivity, as long as they remain aware that they have deliberately left something real off of their maps of nature for the sake of focusing on certain objectifiable and quantifiable aspects. However, along with Bernard Lonergan (1967, 1970), I suggest that a more general empirical method can make room for subjectivity as central to a rich understanding of the natural world. Scientific naturalism, on the other hand, arbitrarily denies that the inner world to which each of us has immediate access is part of nature at all. And by refusing to acknowledge that nature has a pervasive "insideness," naturalists ironically end up espousing implicitly the very dualism they explicitly reject.

The wider kind of "seeing" I am proposing here takes into account both the inside and the outside of things in a single stereoscopic vision of the whole. And by allowing that the natural world is lined with an "insideness" that science itself cannot by definition take into account, one may also suppose that there exists in the universe a capacity to receive and internalize, at least in principle, the purposiveness that religious faith and theology attribute to the whole. In other words, the cosmos may be the carrier of an overarching meaning that will always be inaccessible to a purely objectifying outlook. Nature's insideness, of course, is most immediately accessible in the experience each of us has of our own intelligent functioning. It is only arbitrarily and by decree that naturalists such as Lewontin excise from their formal conception of the universe the fact of subjectivity and purpose.

A richer empiricism, on the other hand, will acknowledge that our own subjectivity belongs to the wider universe as something welling up from within nature and not as an interloper from outside. Alfred North Whitehead, one of the most articulate representatives of the wider empiricism

I am advocating, takes pains to point out that every mental event is fully a part of nature. And yet, as he also rightly asserts, science is "completely dominated by the presupposition that mental functionings are not properly part of nature." This self-restricting aspect of scientific method, he goes on to say "is entirely justifiable, but only if its practitioners remain fully aware of the limitations involved" (Whitehead 1968, 156). Unfortunately, scientific naturalists have generally refused to acknowledge the cognitive limits of science. For this reason, the cosmos of scientific naturalism has been pictured as *essentially* devoid of the subjectivity or insideness that we all experience directly in our mental activity. The universe that naturalism has portrayed as essentially mindless is the philosophical foundation upon which modern naturalistic thought has constructed its understanding of everything, including minds (see also Wallace 2000). An essentially mindless universe, needless to say, is one that is resistant to teleology.

In calling for a more general empiricism, I am emboldened by the work of several major philosophers. I have been following here the ideas of Alfred North Whitehead (1925, 1938, 1968) and Bernard Lonergan (1967, 1970), both of whom make the fact of subjectivity the most important aspect of nature to be taken into account in any adequate worldview. The scientist and philosopher Michael Polanyi also argues persuasively that we cannot understand the cosmos and its evolution unless we realize also that human cognition is something inside and not outside the natural world. Even if science itself is permitted to leave out the fact of subjectivity and the personal dimension of knowing, an adequate philosophy of nature cannot justifiably do so (Polanyi 1967). And as the geologist and religious thinker Teilhard de Chardin announces,

The time has come for us to realize that to be satisfactory, any interpretation of the universe . . . must cover the inside as well as the outside of things—spirit as well as matter. True physics is that which will someday succeed in integrating the totality of the human being into a coherent representation of the world.

Teilhard 1999, 6

The point is that, if the universe *includes* the human, one cannot begin to understand nature in any depth without acknowledging that subjective awareness has evolved in us and that it is, therefore, a terrestrial and, by extension, a cosmic phenomenon (Teilhard 1999, 109ff.).

To anticipate the inevitable objection that what I am proposing here is too anthropocentric, I would emphasize that what I am trying to understand is not ourselves but the *universe*. I am highlighting the fact of subjectivity only because it is usually left out of naturalistic cosmology as though it were not part of the real world. But to leave out subjectivity is to be quite unobjective, since subjectivity is an *objective* aspect of nature. Contrary to Lewontin's *a priori* exclusions, the underlying *experimental spirit* of science, as Teilhard himself suggests, should permit thought to transcend the narrower empiricism of scientific method so as to take into account *all* the data of our experience. A wider empiricism attends first and foremost to the subjectivity that has emerged in living beings as something inside, not outside of, the world of nature (Teilhard 1999, 22–32).

The Possibility of Purpose

Once it is acknowledged that human mental experience is intrinsic to nature and not a ghostly wisp hovering over a hypothetical world machine, the contrived barriers put up by scientific naturalism against the notion of cosmic and evolutionary purpose also begin to fall. Purpose need no longer be understood as something solely "on-the-brain" of misguided religious persons but instead as a real aspect of the universe. Moreover, the purposeful activity of living subjects will then be seen as continuous with, rather than an exception to, what is going on more generally in cosmic process. Indeed, cosmic purpose may be coextensive with, though not reducible to, the long story of nature's gradually intensifying its insideness or subjectivity. The cosmic phenomenon of subjectivity, after all, is one whose initial stages astrophysics finds already being prepared in the earliest moments of cosmic becoming. Of course, there is much more to cosmic purpose than the emergence of subjectivity, but, at the very least, any process that gives rise to increasingly intense modes of experience and subjectivity may plausibly be called purposeful.

How so? Earlier I proposed that any process that is bringing into actuality what is undeniably valuable is purposeful. I think it can reasonably be asserted that our own consciousness is an undeniable instantiation of value. We know this to be true because, even in questioning whether it is true, we are automatically valuing our own minds while we are in the act of raising questions about our mind's value (Lonergan 1967). We cannot help spontaneously prizing our own cognitional apparatus as intrinsically good. And so, if our own mental activity is deeply entangled with the totality of cosmic process—as science itself has now shown to be the case—then an examination of our intelligent subjectivity provides a reliable starting point for thinking about what kind of universe it is that gave rise to our minds.

The bringing about of other living beings and other kinds of subjectivity, of course, also gives purpose to cosmic process. And I have argued elsewhere that the beauty and diversity of nature are indeed intrinsic values that suggest a purposive universe (Haught 2000, 2003). But what sort of meaning or purposiveness can we associate with evolution's three-fold recipe, the one consisting of chance, necessity, and deep time? I believe that the notions of chance and necessity are both lifeless mental abstractions that fail to do justice in any way to the deeper fact that nature is a *story* that blends contingency, law, and time into something truly remarkable and unrepeatable. Since meaning is generally embodied in the form of story, it is conceivable that the irreducibly narrative character of nature opens it up, in a remarkably deep sense, to being the embodiment of purpose. Evolutionary naturalism, unfortunately, isolates chance and necessity from their concrete togetherness in the *story* of evolution. For example, some evolutionary naturalists make chance or contingent events the ultimate explanation of living diversity (Gould and Lewontin 1979). Likewise, ultra-Darwinian adaptationists make the "law" of natural selection the main engine of evolution, and they see the sheer depth of time, with its allowance for a minutely paced gradualism, as a causal "explanation" of evolution (Dawkins 1986, 1995, 1996).

In recent evolutionary naturalism the three ingredients—chance, selection, and deep time—have become mentally segregated, each enshrined at times as a separate kind of explanation of life's diversity. This splintering has caused naturalists to overlook the concrete *togetherness* of contingency, lawful consistency, and deep time as three inseparable aspects of nature's narrative setting. These three natural ingredients are essential for Darwinian evolution, but—much deeper than that—contingency, consistency, and temporality are the stuff of *story*. In the world's dramatic unfolding, there is a contingent openness to indeterminate future outcomes. Then there is the underlying consistency of lawful constraints that limits possibilities so that something determinate can happen and the story can continue. And, of course, there is deep time. In the real world "contingent" openness never exists independently of the habitually constraining and lawful consistency (misnamed "necessity") that gives continuity to nature in its narrative passage through time. Woven into time's depth and irreversibility, nature's contingency and habituality are essential elements in the unfolding of a still-unfinished story.

What requires understanding, therefore, is the delicate *blend* of openness, constraint, and temporality that clothes the cosmos in the apparel of drama. It is this combination—and not chance, necessity, or time considered in mutual isolation—that allows evolution to be purposive. Nature is narrative to the core, and the story is not over. Nature is not a state but a historical genesis, a process of becoming, an epic still being told. And so we shall never get to the bottom of evolution until we have understood why nature is open to narrative in the first place. Contrary to the tenets of naturalism, I doubt that the natural sciences can answer this question without leaving ample room also for theological conjecture at its own appropriate level of understanding.

Life's evolution, let us always keep in mind, is situated within the more foundational context of a *cosmic* story. That the story seems not to be following a rigid plan or that life does not appear to be carefully engineered is nonetheless completely consistent with nature's being a meaningful story. If design ruled everything, as today's "intelligent design" advocates would prefer, necessity and rigidity would have locked life into eternal stasis—death, in other words. There would be no story to tell. Order without novelty is meager monotony. But, blessedly, there exists in nature's and life's contingency an openness to novel possibility that softens up the consistency in natural process. Nature must be open to the future if it is to avoid metamorphosing into hard-rock necessity. Its contingencies and imperfections assist in keeping it fluidly open to the future. To an earlier and now passé brand of Darwinism, it was a theological scandal that many adaptations seemed imperfect, since imperfection spoiled the idea of intelligent divine design. But, as it turns out, the imperfections in organic adaptation are essential if the story is to keep going and remain interesting. If nature is to be a rich narrative, we must remark at how fortunate it is that adaptation and "design" are not comfortably complete. Evolution is much more than the unfolding of algorithmic determinism, as Daniel Dennett has proposed (1995).

On the other hand, openness to transformation does not mean absolute indeterminateness either, as the phenomenon of biological convergence shows (Conway Morris 2003). There is a finite range of possibilities and a channeling aspect to evolution. These keep life from splashing out all over the place in completely unrestrained hyperspace. The patterns assumed by life, whether on Earth or elsewhere, seem to be finite in number. Life is open to possibility, but possibility is not limitless. Otherwise, there would be no continuity or consistency to the story. Evolution arises in a narrative matrix, and narrative requires habituality and redundancy along with novelty to keep the life journey from collapsing at any capricious moment into complete confusion. Contingency, if one still wishes to use this abstract term, adds historicity and dramatic suspense to recurrent natural processes. Ultra-Darwinian naturalism looks for the strain of lawful necessity (and, hence, predictability) in all natural occurrences, and so it is uneasy with contingency as an explanation (Dawkins 1986, 1995, 1996; Dennett 1995). Contingency means uniqueness, singularity, specificity, and unrepeatability; and these all defy the sheer generality and reductive simplicity idealized by scientific naturalism. But pure contingency is no explanation either. When it appears in combination with nature's habituality, contingency is an essential ingredient of story. But when it is absolutized as an independent and ultimate explanation, contingency is equivalent to unintelligibility or absurdity. At the point of being thus maximized, contingency no less than necessity banishes meaning from the world.

I would propose, then, that the naturalistic enshrinement of either chance or necessity can survive only in an illusory and imaginative world of ideas quite cut off from the actual narrative flow of nature and of life itself. And this narrative, a story that wends we know not where, may, for all we know, be pregnant with the promise of ultimate meaning. If so, there may still be abundant room, alongside of science, for a theology of evolution.

Acknowledgments

I would like to extend my thanks to the John Templeton Foundation for the support of the conference at which this paper was first presented. Also thanks are due to George Coyne, S.J., and the Vatican Observatory for hosting the conference and to Simon Conway Morris for leading the discussions and editing the present volume.

References

Atran, Scott. 2002. *In gods we trust: The evolutionary landscape of religion.* New York: Oxford University Press.

Boyer, Pascal. 2001. *Religion explained: The evolutionary origins of religious thought.* New York: Basic Books.

Burkert, Walter. 1996. *Creation of the sacred: Tracks of biology in early religions.* Cambridge, MA: Harvard University Press.

Conway Morris, Simon. 2003. *Life's solution: Inevitable humans in a lonely universe.* New York: Cambridge University Press.

Cziko, Gary. 1995. *Without miracles: Universal selection theory and the second Darwinian revolution.* Cambridge. MA: MIT Press.

Dawkins, Richard. 1986. *The blind watchmaker.* New York: W. W. Norton & Co.

———. 1995. *River out of Eden.* New York: Basic Books.

———. 1996. *Climbing mount improbable.* New York: W. W. Norton & Co.

Dennett, Daniel C. 1995. *Darwin's dangerous idea: Evolution and the meaning of life.* New York: Simon & Schuster.

Flanagan, Owen. 2002. *The problem of the soul: Two visions of mind and how to reconcile them.* New York: Basic Books.

Gould, Stephen Jay. 1977. *Ever since Darwin.* New York: W. W. Norton.

Gould, S. J., and R. C. Lewontin. 1979. The spandrels of San Marco and the panglossian paradigm: A critique of the adaptationist programme. *Proceedings of the Royal Society of London, Series B.* 205, no. 1161: 581–98.

Haught, John F. 2000. *God after Darwin: A theology of evolution.* Boulder, CO: Westview Press.

———. 2003. *Deeper than Darwin: The prospect for religion in the age of evolution.* Boulder, CO: Westview Press.

Klemke, E. D. 1999. Living without appeal: an affirmative philosophy of life. In *The meaning of life,* 2nd edition, 186–97. New York: Oxford University Press.

Lewontin, Richard. 1997. Billions and billions of demons. *New York Review of Books,* January 9.

Lonergan, Bernard, S. J. 1967. Cognitional structure. in *The Collection,* ed. F. E. Crowe, 221–39. New York: Herder and Herder.

———. 1970. *Insight: A study of human understanding.* 3rd. ed. New York: Philosophical Library.

Polanyi, Michael. 1967. *The tacit dimension.* Garden City, NY: Doubleday Anchor Books.

Rose, Michael R. 1998. *Darwin's spectre: Evolutionary biology in the modern world.* Princeton, NJ: Princeton University Press.

Ruse, Michael. 2003. *Darwin and design: Does evolution have a purpose?* Cambridge, MA: Harvard University Press.

Teilhard de Chardin, Pierre. 1962. *Human energy.* Trans. J. M. Cohen. New York: Harvest Books/Harcourt Brace Jovanovich.

———. 1999. *The human phenomenon.* Trans. Sarah Appleton-Weber. Portland, OR: Sussex Academic Press.

Towers, Bernard. 1969. *Concerning Teilhard, and other writings on science and religion.* London: Collins.

Wallace, Alan B. 2000. *The taboo of subjectivity: Toward a new science of consciousness.* New York: Oxford University Press.

Whitehead, Alfred North. 1925. *Science and the modern world.* New York: The Free Press.

———. 1938. *Modes of thought.* New York: The Free Press.

———. 1968. *Process and reality.* Corr. ed. Ed. David Ray Griffin and Donald W. Sherburne. New York: The Free Press.

Wilson, Edward O. 1998. *Consilience: The unity of knowledge.* New York: Knopf.

CONTRIBUTORS

Nicola S. Clayton is professor of comparative cognition in the Department of Experimental Psychology at the University of Cambridge and graduate tutor and fellow at Clare College, Cambridge, U.K. Nicky studies the development and evolution of cognition. Her work is mainly with members of the crow family and comparisons between the crows and apes, including humans. She is currently on the editorial board of *Proceedings of the Royal Society London Series B, Biological Reviews,* and *Public Library of Science One,* consulting editor of *Learning and Behavior,* and an associate editor of *Learning and Motivation.* She has edited one book, *Social Intelligence: From Brain to Culture,* published by Oxford University Press (with Nathan Emery and Chris Frith).

Simon Conway Morris is based in the University of Cambridge and is well known for his work on the Burgess Shale and Cambrian explosion, as well as the study of evolutionary convergence. His most recent book was *Life's Solution* (Cambridge University Press), and he is currently completing his next book, *Darwin's Compass.*

Celia Deane-Drummond has worked for many years at the interface of biological science and theology, having gained doctoral degrees in both plant science and theology. She holds a chair in theology and the biological sciences at the University of Chester, U.K., and is director of the Centre for Religion and the Biosciences that she founded in 2002. She has published numerous books and articles in the field.

Nathan J. Emery is a Royal Society University Research Fellow and senior lecturer in cognitive biology at Queen Mary, University of London as well as an affiliated lecturer at Cambridge University. He is currently working on a long term project "Social reasoning: evolution, cognition and neurobiology" working with members of the crow family as well as also monkeys and apes. He is on the editorial board of "Animal Cognition," "PloS ONE," and "Integrative and Communicative Biology," and has edited two books, *Cognitive Neuroscience of Social Behaviour* (with Alex Easton) and *Social Intelligence: From Brain to Culture* (with Nicky Clayton and Chris Frith).

Robert A. Foley is Leverhulme Professor of Human Evolution at the University of Cambridge and a fellow of King's College, Cambridge, U.K. He has carried out extensive research on human evolution, especially its ecological context and the processes and patterns of the history of our lineage.

Nigel R. Franks is professor of Animal Behaviour and Ecology at the University of Bristol, U.K. His main research fascination is with individual and collective intelligence in ants. He has studied ants professionally for thirty years.

John F. Haught is distinguished research professor in the Department of Theology at Georgetown University. His area of specialization is systematic theology, with a particular interest in issues pertaining to science, cosmology, evolution, ecology, and religion. In 2002, he was the winner of the Owen Garrigan Award in Science and Religion, and in 2004, the Sophia Award for Theological Excellence. He has authored many books, articles, and reviews and lectures internationally on issues related to science and religion.

Richard E. Lenski is the John A. Hannah Distinguished Professor of Microbial Ecology at Michigan State University. His research program emphasizes experimental approaches to testing evolutionary ideas. He has had fellowships from the Guggenheim and MacArthur Foundations and been elected to the American Academy of Arts and Sciences and the National Academy of Sciences.

George McGhee is professor of Geological Sciences, Ecology, and Evolution at Rutgers University, and he is a past fellow of the Konrad Lorenz Institute for Evolution and Cognition Research in Vienna, Austria. His latest book is *The Geometry of Evolution,* published by Cambridge University Press.

Karl J. Niklas joined the Cornell University faculty in 1978, where he is the current Liberty Hyde Bailey Professor of Plant Biology. He is the author of over 255 scientific articles and three books (*Plant Biomechanics, Plant Allometry,* and *The Evolutionary Biology of Plants*).

Michael Ruse is professor of philosophy at Florida State University. He is the author of a number of books, most recently *Darwinism and Its Discontents,* published by Cambridge University Press.

Anthony Trewavas is a plant biologist with specialist interests in plant-cell signal transduction and plant behavior. He has published over 220 papers and two books. He is a fellow of the Royal Society of London, Royal Society of Edinburgh, and Academia Europea; has been elected to life membership of the American Society of Plant Biologists; and is an original member of the Institute for Scientific Information's (ISI) most highly cited researcher group.

Hal Whitehead's research is focused on social and cultural evolution, particularly among whales. He spends periods at sea on sailing boats following whales and collecting data. He develops methods for studying social structure and culture in animals and builds computer models of cultural evolution.

INDEX